Introduction to Digital Audio
Pro Tools Version Second Edition

Introduction to Digital Audio

Tony M Dofat

TDC Group, Inc.
New York, NY
TDC
GROUP
2016

First Printing: 2016

ISBN 978-0-578-17904-9

TDC Group, Inc.

www.thetdcgroupinc com

For special orders, please call (646) 402-5519

Contact: tony@thetdcgroupinc.com

Dedication

I dedicate this book to my students and the Media Arts Department at Long Island University.

Contents

Acknowledgements

I'd like to acknowledge the Audio Engineers and Mixers that inspired me to become better at my craft. I'd like to thank Avid, Apple, and Akai.

(This page was intentionally left blank)

Preface

This book is for the beginner and considered an introduction to the DAW (Digital Audio Workstation) and digital world. I find it extremely relevant to explain, in depth, the science behind digital audio in conjunction with applying the digital audio technology with our music. You will learn the science behind digital technology on an introductory level and how or why the digital workstation has surpassed and replaced multi-million dollar recording facilities. Digital Audio technology has been around for almost 50 years and has become the industry standard today. You will also learn about A/D conversion, bit depth, Pro-tools 12 functions and features, importing, exporting, internal routing, sends, and inserts, along with plug-in usage and synchronizing digital video with audio tracks within your DAW.

The concept of this book was created around a syllabus with some additions and revisions from myself as I instructed a Digital Audio for a University. As we know, software changes quite often as we are now operating Pro-Tools 12 but the technology of digital audio has been consistent for many years. This book is for the beginner and considered an introduction to the DAW and digital world.

Introduction

Whether laying in bed, riding the train, or working out and listening to your favorite song through a pair of $300 headphones and feeling every beat down to the sample, not realizing the development and advancement of analog technology to the digital world and also not old enough to experience the audio struggles of the analog era. Why does the quality of music sound 10x cleaner than the 70's and 80's? There are many things that plays a factor in today's music sonically and the most important and most crucial part is the standardized workstation or DAW (Digital Audio Workstation) in which was used to record, edit, and mix that composition, Pro Tools.

This book will teach you the methods behind digital technology on an introductory level and how or why the digital workstation has surpassed and replaced multi million dollar-recording facilities. I am somewhat amazed that the invention of digital technology was in the 70's and here we are almost 50 years later, digital technology is finally at its highest peak. You will learn A/D conversion along with the signal flow of your basic DAW MIDI set-up.

We will be using Pro Tools version 12 as part of our explanation and learning process of digital audio. Once you
Learn the fundamentals and processing of digital Audio technology, you will master any DAW.

2

(This page was intentionally left blank)

Chapter 1

What Digital

Digital describes electronic technology that generates, stores, and processes data in terms of two states: positive and non-positive. Positive is expressed or represented by the number 1 and non-positive by the number 0. Thus, data transmitted or stored with digital technology is expressed as a string of 0's and 1's. Each of these state digits is referred to as a bit (and a string of bits that a computer can address individually as a group is a byte). Prior to digital technology, electronic transmission was limited to analog technology, which conveys data as electronic signals of varying frequency or amplitude that are added to carrier waves of a given frequency. Broadcast and phone transmission has conventionally used analog technology.

Digital technology is primarily used with new physical communications media, such as satellite and fiber optic transmission. A modem is used to convert the digital information in your computer to analog signals for your phone line and to convert analog phone signals to digital information for your computer. Digital describes any system based on discontinuous data or events. Computers are digital machines because at their most basic level they can distinguish between just two values, 0 and 1, or off and on. There is no simple way to represent all the values in between, such as 0.25. All data that a computer processes must be encoded digitally, as a series of zeroes and ones. The opposite of digital is analog, for example; a clock in which the hands move continuously around the face. Such a clock is capable of indicating every possible time of day. In contrast, a digital clock is capable of representing only a finite number of times (every tenth of a second, for example).

In general, humans experience the world analogically. Vision, for example, is an analog experience because we perceive infinitely smooth gradations of shapes and colors. Most analog events, however, can be simulated digitally. Photographs in newspapers, for instance, consist of an array of dots that are either black or white. From afar, the viewer does not see the dots (the digital form), but only lines and shading, which appear to be continuous. Although digital representations are approximations of analog events, they are useful because they are relatively easy to store and manipulate electronically. The trick is in converting from analog to digital, and back again. This is the principle behind compact discs (CDs). The music itself exists in an analog form, as waves in the air, but these sounds are then translated into a digital form that is encoded onto the disk. When you play a compact disc, the CD player reads the digital data, translates it back into its original analog form, and sends it to the amplifier and eventually the speakers.

Digital Audio is a technology in which an audio signal has been converted, replicated, and emulated to the exact shape and sound of an analog waveform by the method of sampling, thousands of times in very small increments per second. Analog audio is converted to a digital format by using a converter or ADC interface; which stands for Analog to Digital converter, also known as an A/D Interface. Analog sound waves are converted to digital for recording and editing purposes within your Digital Audio Workstation otherwise known as a DAW for short. Once your analog files are converted into digital, they will have to be converted back to analog by using the same ADC interface, which is also simultaneously a digital to analog converter.

A digital audio system starts with an ADC that converts an analog signal to a digital signal. The ADC runs at a specified sampling rate and converts at a known bit resolution. CD audio, for example, has a sampling rate of 44.1 kHz (44,100 samples per second), and has 16-bit resolution for each stereo channel. Analog signals that have not already been band limited must be passed through an anti-aliasing filter before conversion, to prevent the distortion that is caused by audio signals with frequencies higher than the Nyquist frequency (the minimum rate at which a signal can be sampled without introducing errors, which is twice the highest frequency present in the signal).

A digital audio signal may be stored or transmitted. Digital audio can be stored on a CD, a digital audio player, a hard drive, a USB flash drive, or any other digital data storage device. The Digital signal may then be altered through digital signal processing (DSP), where it may be filtered or have effects applied. Audio data compression techniques, such as MP3 (MPEG-2 Audio Layer III), Advanced Audio Coding (AAC), Ogg Vorbis, or FLAC, are commonly employed to reduce the file size. Digital audio can be streamed to other devices. For playback, digital audio must be converted back to an analog signal with a DAC (ADC Interface). DACs run at a specific sampling rate and bit resolution, but may use oversampling, upsampling, or downsampling to convert signals that have been encoded with a different sampling rate.

Digital audio technologies today are used in the recording, manipulation, mass-production, and distribution of sound, including recordings of songs, instrumental pieces, podcasts, sound effects, and other sounds. Modern online music distribution depends on digital recording and data compression. The availability of music as data files, rather than as physical objects, has significantly reduced the costs of distribution. Before digital audio, the music industry distributed and sold music by selling physical copies of albums in the form of records, tapes, and then CDs. With digital audio and online distribution systems such as iTunes, companies sell digital sound files to consumers, which the consumer receives over the Internet. This digital audio/Internet distribution model is much less expensive than producing physical copies of recordings, packaging them and shipping them to stores.

An analog audio system captures sounds, and converts their physical waveforms into electrical representations of those waveforms by use of a transducer, such as a microphone. The sounds are then stored, as on tape, or transmitted. The process is reversed for playback: the audio signal is amplified and then converted back into physical waveforms via a loudspeaker. Analog audio retains its fundamental wavelike characteristics throughout its storage, transformation, duplication, and amplification. Analog audio signals are susceptible to noise and distortion, due to the innate characteristics of electronic circuits and associated devices. Disturbances in a digital system do not result in error unless the disturbance is so large as to result in a symbol being misinterpreted as another symbol or disturb the sequence of symbols. It is therefore generally possible to have an entirely error-free digital audio system in which no noise or distortion is introduced between conversion to digital format, and conversion back to analog.

A digital audio signal may be encoded for correction of any errors that might occur in the storage or transmission of the signal, but this is not strictly part of the digital audio process. This technique, known as "channel coding", is essential for broadcast or recorded digital systems to maintain bit accuracy. The discrete time and level of the binary signal allow a decoder to recreate the analog signal upon replay. Eight to Fourteen Bit Modulation is a channel code used in the audio Compact Disc (CD).

Up to the 1970s, all recording technology depended on creating a physical analog, whether on tape or disk, of the original sound waves. Despite many incremental improvements to these techniques, by the 1970s, further reductions in noise and distortion were becoming increasingly difficult and expensive. As a result, audio researchers began to experiment with digital techniques that had first been exploited in the computer and telecommunication industries. Digitizing an electrical audio signal consisted of first sampling the wave thousands of times a second, measuring the amplitude of each sample, and then assigning one of a limited number of binary values to each. The resulting tape recording consisted of a coded series of on-off signals, or bits. Unlike analogue recording, in which noise and distortion tended to accumulate at each stage of recording and post-production, in digital recording the original message could be cleanly regenerated as long as the simple binary values were recognizable. In addition, minute alterations to the message could be made electronically, by altering the value of individual bits.

The first digital tape recorder was demonstrated in Japan in 1967, with the first digitally mastered records appearing on the Denon label in 1972. The first commercially available digital audio recorder was the Sony PCM-1. Introduced in 1977, the PCM-1 (Pulse-code modulation) converted an incoming analog signal into a digital one, which was then recorded onto a standard videocassette in a VCR.

Pulse-Code Modulation (PCM) was invented by British scientist Alec Reeves in 1937 and was used in telecommunications applications long before its first use in commercial broadcast and recording. Commercial digital recording was pioneered in Japan by NKH and Nippon Columbia, also known as Denon in the 1960s. The first commercial digital recordings were released in 1971. The BBC also began to experiment with digital audio in the 1960s. By the early 1970s it had developed a 2-channel recorder, and in 1972 it deployed a digital audio transmission system that linked their broadcast center to their remote transmitters. The first 16-bit PCM (Pulse-Code Modulation) recording in the United States was made by Thomas Stockham at the Santa Fe Opera in 1976 on a Soundstream recorder. An improved version of the Soundstream system was used to produce several classical recordings by Telarc (Independent record label) in 1978. The 3M Digital Multitrack Recorder in development at the time was based on BBC technology and the first all-digital album recorded on this machine was in 1979.

In the early 1980's, Sony and Mitsubishi developed some Popular digital multitrack recorders which aided in making digital recording's somewhat accepted by major record companies and In 1982, the introduction of the CD (Compact Disc) popularized digital audio with consumers.

Initially, digitally mastered records were still released on vinyl disks in the analog format. In 1982, however, Sony and Philips released the first compact discs and players. On a CD, the digital information was embodied as millions of microscopic bits in the reflective aluminum coating of the disk. The CD player employed an optical unit to "read" the pattern of bits and convert the resulting electrical pulses into an analogue signal that could drive a speaker. Because the system was optical, friction and noise from physical contact between stylus and record was eliminated. In 1987, another digital format was introduced, the digital Audio Tape (DAT). Due to record company opposition to a medium that allowed flawless copies of CDs to be made, few DAT recorders were released to the North American public. They did, however, become quite common for professional recording.

Digital recording devices are continually evolving. In the last decade, the most significant, and largely unforeseen, development has been the widespread exchange of digital music files over the Internet. This has been made possible by a number of technical developments, including software for "ripping" songs from commercially released CDs, compression software that reduces the size of these music files by eliminating redundant or unnecessary data (e.g. MP3), and file-sharing software that supports the swapping of these files over the Internet (such as Napster, Kazaa and LimeWire). The popularity of MP3 file sharing, especially among young, technically savvy music lovers with disposable incomes, soon encouraged manufacturers to introduce portable MP3 players. The first of these, the Elger Labs

MPMan F10 and the Diamond Rio PMP300 were introduced in 1998. These employed flash memory chips for data storage and had a capacity limited to less than a dozen songs. In late 1999, Remote Solutions introduced the first MP3 player incorporating a magnetic hard-drive, which boasted a capacity of 1,200 songs. This was followed two years later by Apple Computers' iPod. The huge popularity of file sharing has shaken the foundations of the recording industry, whose profit for over a century has depended on restricting the ability of record buyers to make and transmit high-quality, free copies of their products. It is not yet clear what arrangements will ultimately be devised to balance the rights of music creators to compensation with the rights of consumers to reasonable copying and sharing of their products.

Below are illustrations explaining how analog signals are emulated by digital technology.

This is an analog waveform

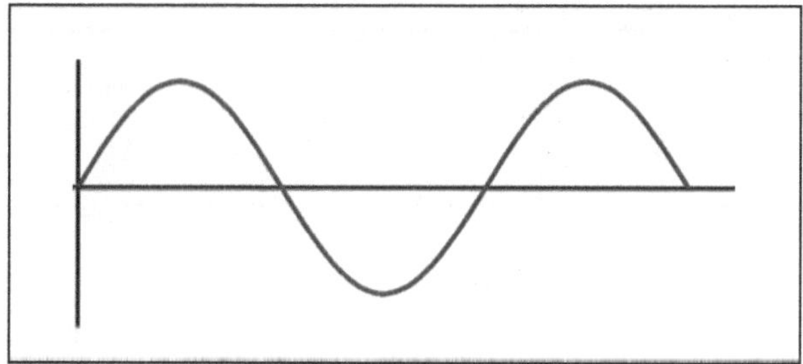

Digital replicates the analog sound wave represented
By the dots, which are samples.

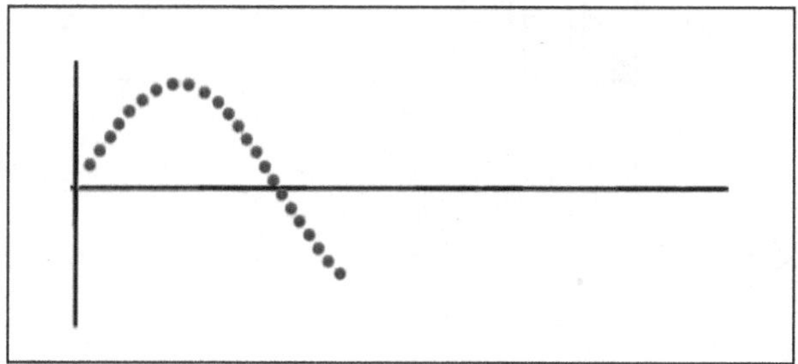

Samples are measured by 1-second intervals. CD quality has 44,100 samples per second.

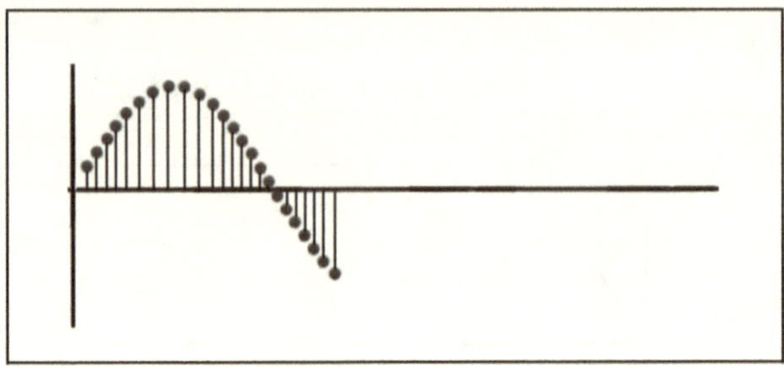

The more samples there are per second, the more accurate and smooth the waveform becomes.

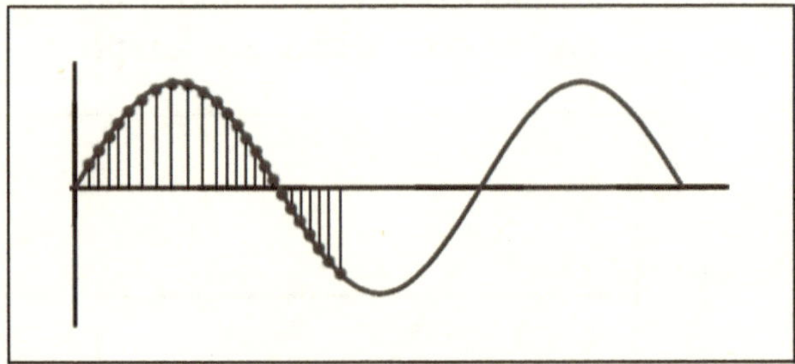

Sample rates vary, depending on your DAW but Pro-Tools HD samples from 44.1kHz, 48kHz, 88.2kHz, 96kHz, 176.4kHz, and 192kHz. After the sample rate goes beyond 44.1, it's very hard for the human ear to detect any differences, higher sample rates will give the engineer a more precise edit thus the reason to use higher sample rates.

The bit depths of each sample also play an important role in the sound and quality of the analog to digital conversion. Bit depths are usually 16, 24, 32, and 64 bits and CD quality is 16 bit. The bit depth is the quality of each sample by providing information, or bits per sample so a lower bit rate will producer a lower quality sample which would result in a lower quality sound which your ears could definitely hear the difference from a lower bit rate as opposed to a higher bit rate.

Chapter 2

DAW

Prior to the digital era, Producer and composers were subjected to create music by using a rig which consisted of multiple rack sound modules, keyboards, drum machines, and sequencers which were all connected via MIDI. Musical Instrument Digital Interface is a protocol developed in the 1980's; which allows electronic instruments and other digital musical tools to communicate with each other. MIDI itself does not make sound, it is just a series of messages like "note on," "note off," "note/pitch," "pitch bend," and many more. These messages are interpreted by a MIDI instrument to produce the sound.

A MIDI instrument can be a piece of hardware (electronic keyboard, synthesizer) or part of a software environment (Reason, FL Studio, and logic Pro). The advantages of MIDI include(s): An entire song can be stored within a few hundred MIDI messages (compared to audio data which is sampled thousands of times a second) easy to modify/manipulate notes -change pitch, duration, and other parameters without having to rerecord change instruments. Remember, MIDI only describes which notes to play; you can send these notes to any instrument to change the overall sound of the composition.

The rear of the MIDI racks would be a massive amount of quarter inch audio and MIDI cables intertwined and tangled like a mess. 20 years later, we are at the peak of the digital audio workstation, or DAW software, which replaces the once used massive MIDI rig along with your 10 pound drum machines, everything is now contained within your computer or laptop, how convenient is that?

The digital audio workstation computer software does more than just allow you to compose a track, the DAW's main function is to provide every need of the producer or composer which includes vocal recording, non-linear non-destructive recording and editing, extremely high sample rates and bit resolutions up to 192,000 samples per se-second and bit resolutions up to 64 bit, and the luxury of packing your entire DAW inside your laptop bag. In digital video editing, non-linear editing is a method that allows you to access any frame in a digital video clip regardless of sequence in the clip. The freedom to access any frame, and use a cut-and-paste method, similar to the ease of cutting and pasting text in a word processor, and allows you to easily include fades, transitions, and other effects that cannot be achieved with linear editing.

In the early days of electronic video production, linear (tape-to-tape) editing was the only way to edit. In the 1990s, non-linear editing computers became available and opened a whole new world of editing power and flexibility. Everyone did not welcome non-linear editing and many editors resisted the new wave. In addition, early digital audio

Was plagued with performance issues and uncertainty. However, the advantages of non-linear recording and editing eventually became so overwhelming that they could not be ignored.

In the 21st Century non-linear gained dominance and linear editing headed towards obsolescence and today, it's completely obsolete. During this time the description "non-linear" was slowly abandoned as it was no longer necessary, almost all editing was now digital and the "non-linear" aspect was assumed, linear was dead.

The following are some popular DAWs of today

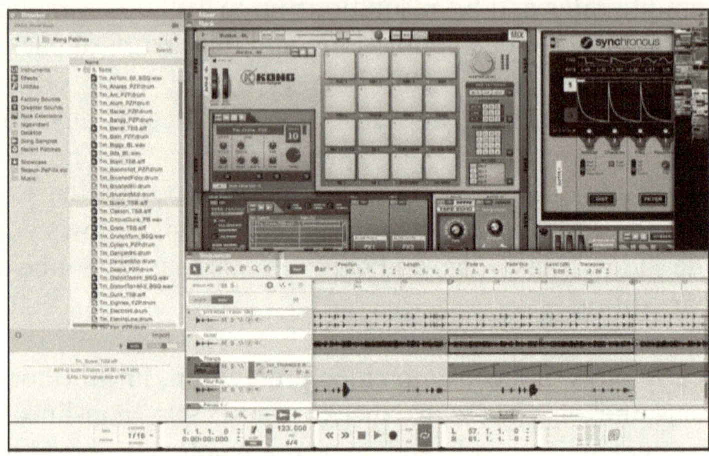

Propellerhead Reason 8

The latest version, released in 2015, Reason 8, introduces several enhancements that make this DAW a more modern one, easier to use and manage. Some buttons and controls were changed, now being much more intuitive, clean and logical to use. Available in two versions, the main version with all instruments, effects and all features included and the "Essentials" version unfortunately has some limitations.

Apple LogicPro x

Apple's Logic Pro X is a Mac-only music software, very powerful and affordable with strong production capabilities
Logic Pro X 10.2 brings even more upgrades and more obvious is the addition of Alchemy, the next-generation sample manipulation synthesizer with over 3000 presets ready for all types of electronic music. Besides Alchemy synth, there are more features such as the Logic Remote from where you can control your DAW from any corner of your studio using an iPad.

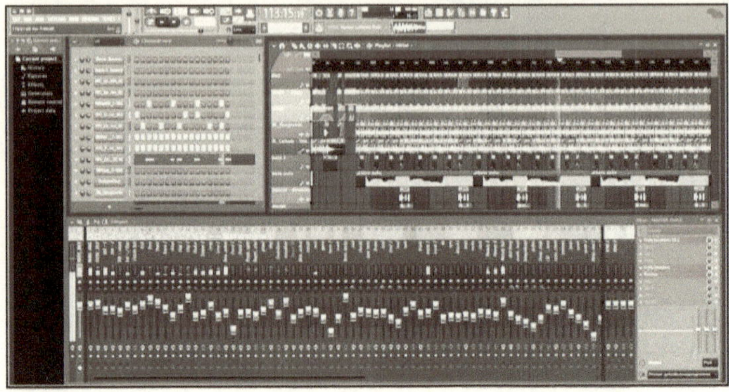

FL Studio 12

One of the best music software used and appreciated by many electronic music producers across the globe. Beat makers such as Lex Luger, Shawty Red, Young Chop, Jahlil Beats, and Hit Boy uses this DAW. Also, Jay-Z and Kayne's hit record, "N***** in Paris" was produced with FL Studio. So, if you are looking for a fun and easy-to-use DAW, with a graphical interface that looks great and can be customized according to your preferences, then FL Studio 12 is a very good choice.

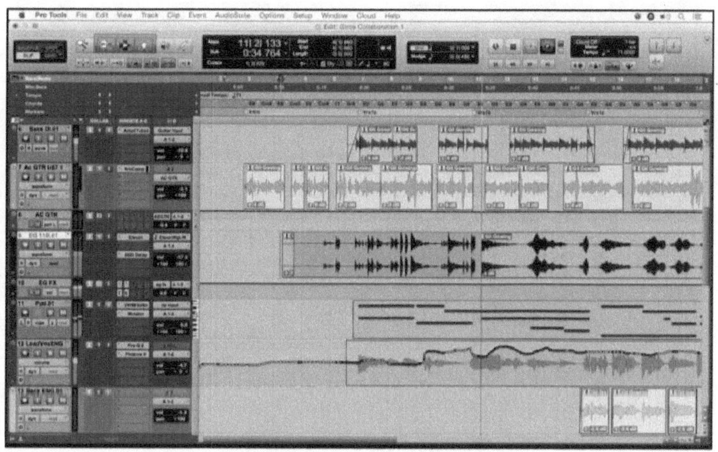

Avid Pro-Tools 12

Pro Tools 12 is the next generation of the industry-standard music making audio software with artist collaboration support via Avid Cloud Collaboration. Pro Tools v12.1 updates and improvements for Editing, AudioSuite, Bounce, Audio, and Video Engines, Delay Compensation, I/O Setup, MIDI, and Satellite workflows to further improve stability.

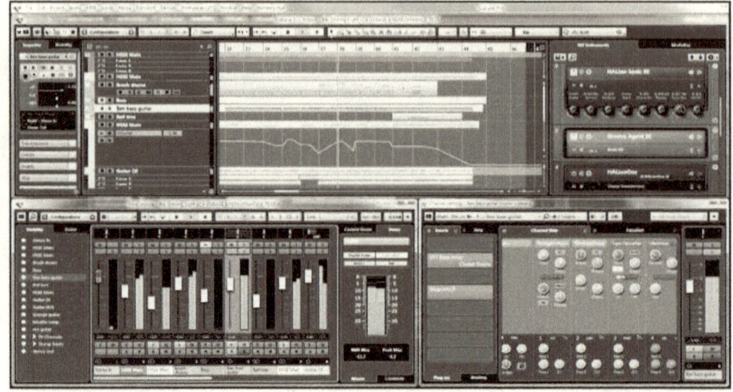

Steinberg Cubase 8

Cubase 8 developed by Steinberg is their latest and greatest version of Cubase music software with many new features and comes in two versions: Artist and Pro. Cubase Pro has all the features of Artist, plus a lot of features such as a Full Scoring Editor, VariAudio for pitch correction, Surround Sound, VCA Faders, Direct Routing, Wave View, significant performance increase and many GUI refinements.

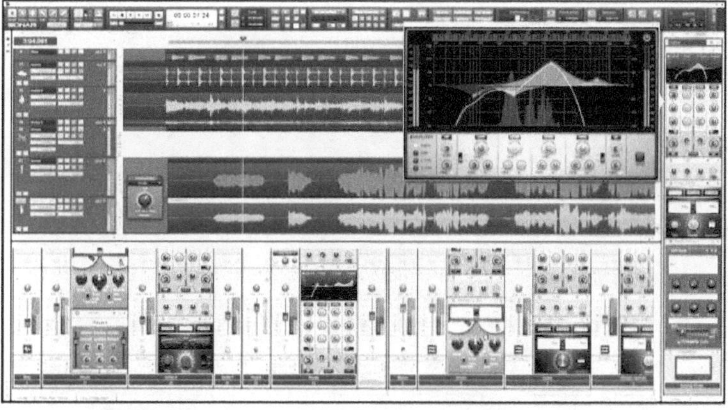

Cakewalk Sonar

One of the more advanced digital audio workstation for Windows (PC) available today. Sonar 2015 comes with a brand-new interface ("Skylight"), mix recall modes, vocal sync, Melodyne integration, and some great new useful tools with new effects. Also, SONAR (Platinum, Professional) comes with Addictive Drums 2 (a great drum virtual instrument) included.

Ableton Live 9

Ableton Live 9 is a clip-based sequencer that can be used for both studio-inside music production and live performances. With a better audio warping engine, a great creative work-flow, and MIDI editing refinements, this DAW offers more flexibility than ever. The advanced automation abilities such as adding curves to automation or duplicating certain automation plus a full range of effects and instruments are ready to turn any idea into reality.

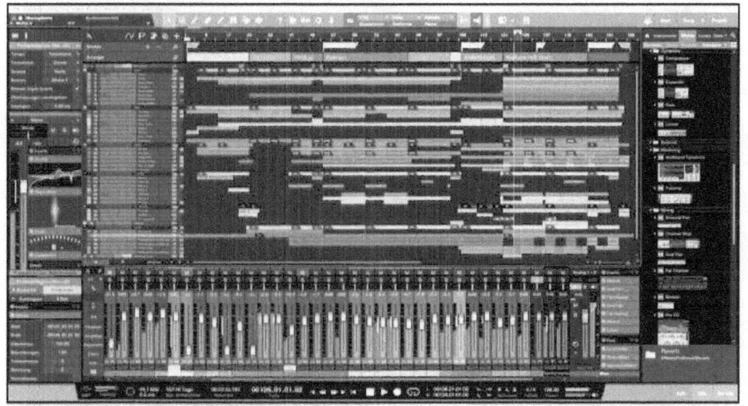

Presonus Studio One 3

Studio One v3 developed by PreSonus is a very solid update, bringing some exciting new features such as the Scratch Pads and the Arranger Track, offering more flexibility than ever. Besides all this, Studio One 3 brings some long-awaited virtual instruments add-ons such as Mai Tai, a brand new polyphonic analog modeling synthesizer and the Presence XT sampler, an enhanced version of the previous Presence sample.

Reaper 5 by Cockos

Reaper is a brilliant piece of music software providing all tools you need to make the best music. Reaper is a multi-track audio and MIDI recording, mixing and mastering, editing and processing digital workstation for both Windows and Mac OS. Reaper 5 was released in 2015 and brings many improvements such as FX VST3 support, video support and performance, new GUI layouts, new automation options, new theme, and much more.

The most popular and the most standard DAW in 100% of all recording studios is Avid Pro-tools; which is also the most professional and most high-end Digital workstation. Pro-Tools HDX system will allow you to sample up to 192kHz per second and 64 Bit resolution. The Quality and DSP (Digital Signal Processing) power is top notch and nothing comes close to the stability of Pro-Tools. If CD quality is 44.1kHz and 16 bit, why record at 192kHz and 32 or 64 bit, meanwhile your ears will not be able to tell the difference because CD quality is somewhat the best that exists. The engineer will chose to record at higher sample rates, not for the sound but for the accuracy of the editing.

One of the things which makes a DAW easier to work with, is the compatibility of using a control surface along with it. A control surface is a human interface device (HID) which; allows the user to control a digital audio workstation or other digital audio application. Generally, a control surface will contain one or more controls that can be assigned to parameters in the software, allowing tactile control of the software. Control surfaces are connected via USB.

USB is short for Universal Serial Bus is an industry standard developed in the mid-1990s that defines the cables, connectors and communications protocols used in a bus for connection, communication, and power supply between computers and electronic devices. USB was designed to standardize the connection of computer peripherals (including keyboards, pointing devices, digital cameras, printers, portable media players, disk drives and network adapters) to personal computers, both to communicate and to supply electric power. It has become commonplace on other devices, such as smartphones, PDAs and video game console and has Effectively replaced a variety of earlier interfaces, such as serial and parallel ports, as well as separate power chargers for portable devices.

Unlike the traditional analog recording, DAWs have eliminated the need to carry the old 35-pound analog 2" Ampex tapes from studio to studio and also eliminating the need to print SMPTE time code and calibrating or aligning the tape machines. The best feature of recording digitally is the "undo" feature. This feature has saved many hit songs from accidental recording or deleting unlike recording analog. Once you record over an audio piece on an analog tape machine, consider it goodbye and gone forever.

Chapter 3

Signal Flow

Signal flow is very important whether you're recording by way of analog or digital. Signal flow is the path in which the audio flows throughout your dynamic chain from the source to the output, meaning speakers or headphones. From your vocalist's voice, the sounds would travel through the microphone, microphone preamplifier, Compressor, Equalizer, ADC or Converter interface, Computer that contains your DAW, Digital to Analog converter, and then finally your speakers or headphones would be the final destination. Below is a detailed explanation regarding each piece of equipment within the chain.

- **Microphone:** a device used in sound-reproduction systems for converting sound into electrical energy, usually by means of a ribbon or diaphragm set into motion by the sound waves. For studio recording purposes, there are three types of microphones, a dynamic, condenser, and ribbon.

I. Dynamic Microphone

One of the most basic differences between microphones is their operating principle; we'll look at Both, identifying their strengths and weaknesses. The word dynamic sounds like something straight from an infomercial, doesn't it? While there are several definitions for the word, the one we're going with is: "have or relating to variation of intensity, as in musical sound." Microphones in general are a specific kind of electromechanical device called a transducer, converting mechanical energy into electrical energy. In the case of a dynamic microphone, a very thin diaphragm of Mylar or other material is attached to a coil of hair-thin copper wire. The coil is suspended in a magnetic field and, when sound vibrates the diaphragm, the coil moves up and down, creating a very small electrical current. The electrical signal is attached to a connector on the microphone. Just plug it in and it works.

Dynamic microphones come in many shapes and sizes, some with large diaphragms, others small. The size and design of the dynamic element plays a big role in how the mic sounds. In general, the larger the diaphragm, the smoother and deeper the sound. For instance, the venerable large-diaphragm "Electro Voice RE-20" is prized for its silky smooth vocal reproduction. It is also a favorite for brass instruments and bass drums. Another iconic dynamic microphone is the Shure SM-58. Virtually unchanged since its introduction in 1966, the SM-58 is a favorite vocal microphone in live sound and concert applications. Using the same dynamic element in a different package, the SM-57 has its own place in history. In addition to duty as the de facto snare drum microphone, you can see the SM-57 in action every time the President of the United States speaks. His podium always mount two SM-57s complete with windscreens and a Dual-microphone mount. Why does the White House choose a $100 dynamic microphone when there are so many other options? In short, the SM-57 is chosen for its durability, reliability and consistent sound quality in virtually every situation. This is generally true for all dynamic mics. Since there are very few parts to break and no electronics to fail, dynamic microphones survive all kinds of torture from the road and the elements. It's not unusual for a professional dynamic mic to take all kinds of abuse and keep working for decades. If there were a downside to dynamics, it would be sensitivity and frequency range, although both shortcomings have been minimized over the years with new materials and magnets.

II. Condenser Microphone

Condenser microphones use a different type of transducer. Condenser elements are essentially a fancy kind of capacitor. In most designs, there is a fixed back plate and a movable front plate. These plates are fixed a specific distance apart and an electrical charge is applied. As sound hits the movable plate, it flexes or vibrates. This creates a tiny change in the capacitance, which can be turned into an electrical signal. However, this signal is extremely small and needs some amplification before it is suitable for connection to audio equipment. Consequently, condenser microphones require a fair amount of electronic circuitry to produce recordable sound. The electronics package needs some type of power source, which is supplied through the sound console, an external power supply or even batteries.

Condenser microphones are also available in a wide range of shapes and sizes - from the size of a pencil eraser to that of a bottle of your favorite soft drink. They tend to be more sensitive and accurate than their dynamic cousins and are found in every area of audio from communications to recording studios. Your cell phone, computer and even some cars have small condenser microphones built in. You'll find them embedded in your camcorder, mounted on microphone stands and suspended from boom poles. They're everywhere! One of the most famous studio condenser microphones is the Neumann U-87. Known for its sparkling clean sound on everything from drums and pianos to vocals, the U-87 is a pricey but desirable addition to anyone's microphone collection. In recent years, Chinese manufacturers have dominated the cost-effective condenser microphone arena. Just check the pages of any music catalog, and you'll find studio-type condenser microphones starting under only $100. There are also many very nice microphones in the $300-500 range. Of course, it's rare for one of these cheaper microphones to rival one with a $3000 ones, but the technology has evolved to the point that anyone can afford a nice condenser microphone.

Many times, microphone selection is based purely on the application. If you need a lapel or shotgun mic, you're looking for a condenser. It's virtually impossible to manufacture a dynamic mic in those forms and, since condensers aren't really dependent on size, they are the perfect choice for the job. On the other hand, if you're using microphones in crazy weather conditions or dirty, unprotected circumstances, you may want a dynamic model instead. In addition, there are microphones suited to particular tasks. For instance, there are dozens of Condenser microphones produced specifically for use in theater and church applications. Whether hanging from a ceiling, mounted on the floor or even a podium, these microphones are designed for optimum pickup of voices in this unique environment, and there are no dynamic equivalents. On the other hand, a band touring the country might opt for more durable dynamic microphones, especially for vocals, guitars, and drums.

III. Ribbon Microphone

Nearly a century old and considered old-fashioned and fin-icky, Ribbon microphones were the first high-quality microphones available. The first commercial ribbon microphone was introduced in the late 20's, the RCA44A. The ribbon microphone is a type of dynamic microphone that uses a very thin metal ribbon suspended in a magnetic field. Sound waves cause the ribbon to vibrate, creating an electrical signal.

This is similar to the way in which a moving-coil microphones works, in fact, they're both dynamic microphones. However, the clever bit is that the ribbon itself acts as both the diaphragm and the conductor. Compared to the stiffer, heavier moving-coil arrangement, a ribbon is ultra-light and has far more freedom of movement. There is therefore greater potential for accurate frequency and transient responses.

Even so, there's more to making a ribbon microphone than lashing a strip of tin foil between two fridge magnets. The ribbon has to be loose enough to move backwards and forwards freely in response to air pressure. Sideways and twisting movements are undesirable, so most manufacturers corrugate their ribbons at either end, or along the entire length, to increase stiffness. The sound of a ribbon microphone is mellow and rounded, warm, sound with a gently rolled off top end that contrasts dramatically with the forward sound of most dynamics and the crystalline, trebly sound of condensers. Nothing sounds quite like a ribbon microphone and also very delicate. Even a puff of breath into an unshielded ribbon can ruin it, and re-ribboning a microphone can be costly so pop filters are very essential. Most old ribbon microphones put out very low-level signals, requiring preamplifiers with a lot of clean gain and the right sort of input impedance. A decent ribbon microphone could cover all bases, combining the best qualities of both microphones in one convenient package.

- **Microphone Pre Amplifier**

A microphone preamplifier is a sound engineering device that prepares a microphone signal to be processed by other equipment. Microphone signals are often too weak to be transmitted to units such as mixing consoles and recording devices with adequate quality. Preamplifiers increase a microphone signal to line level (i.e. the level of signal strength required by such devices) by providing stable gain while preventing induced noise that would otherwise distort the signal.

- **Compressor**

Compression is an electronic effect unit that reduces the volume of loud sounds or amplifies quiet sounds by narrowing or "compressing" an audio signal's dynamic range.

- **Equalizer**

A device used to correct the frequencies of audio signals

- **ADC**

An analog to digital converter is an interface that is connected to your computer via USB or Fire wire and used to convert analog signals into digital information, which is a mandatory process in order for your DAW to record or edit the signal. This equipment works in both directions as it also converts digital to analog signals so your speakers or headphones will be able to play back your signals or sounds.

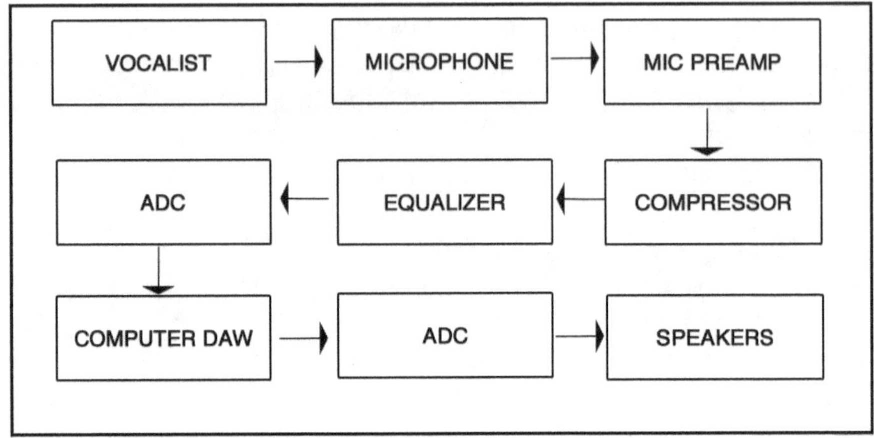

This is your standard professional signal flow chart designed for a dynamic microphone chain to process your vocals prior to recording into you DAW.

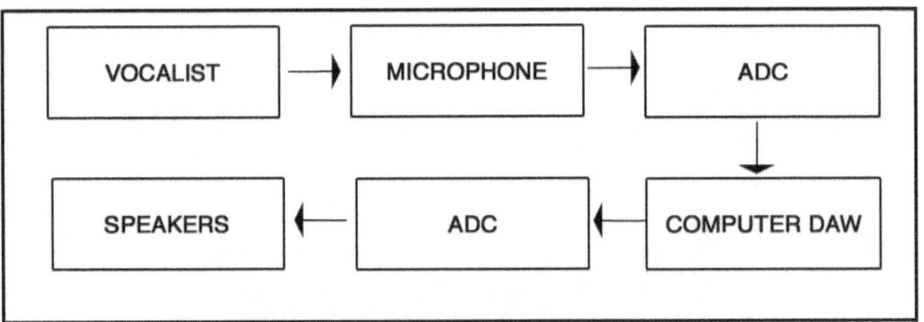

This is the most economical home recording signal flow method for recording directly into your DAW

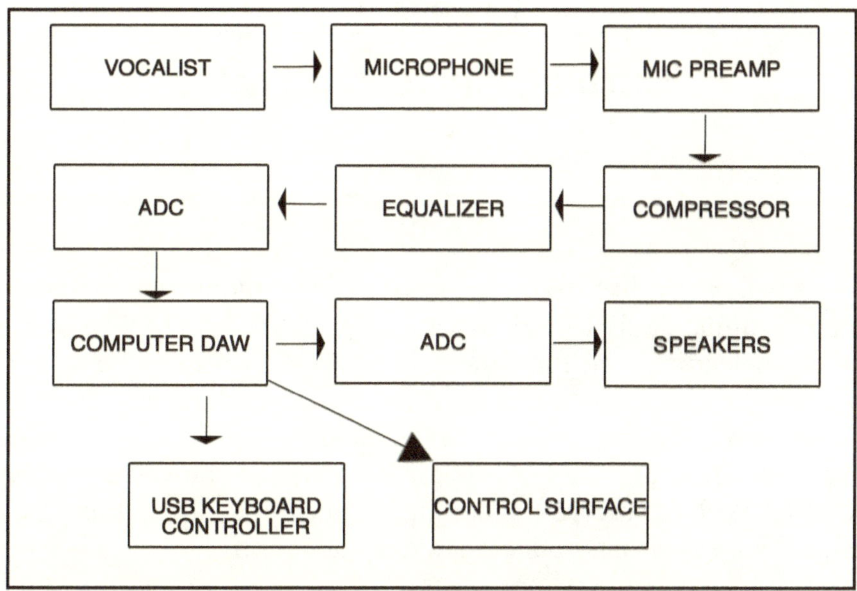

This is your standard professional signal flow chart designed for a dynamic microphone chain to process your vocals prior to recording into you DAW but also including a Keyboard controller for creating and a control surface for mixing.

Chapter 4

Introduction to Pro Tools

Now that we've discussed digital technology and the comparison of analog to digital audio as well as various professional DAW's, we will now begin learning the fundamentals of Avid Pro-Tools version 12. Pro-Tools can get very confusion if we decide to dig deep into every function and parameter without understanding or learning the basics so keep in mind that this is just an introduction, however you will learn a lot such as; recording, editing, importing audio and importing session data, exporting, bouncing to disk, and more.

As with any DAW or computer program, my first rule is to make sure to always save your session after every move or revision. Pro-tools is equipped with auto save within the preference panel but manually saving is your best bet. If you decide to use the auto save, you can set the parameters to save at any interval of time, If you are recording on an unstable computer, it's best to set your auto save for every minute or so.

Let's start by opening the program Pro-Tools 12 by clicking on the icon. The program should load within a minute or so, after it fully loads, a dialog box will appear similar to the following image.

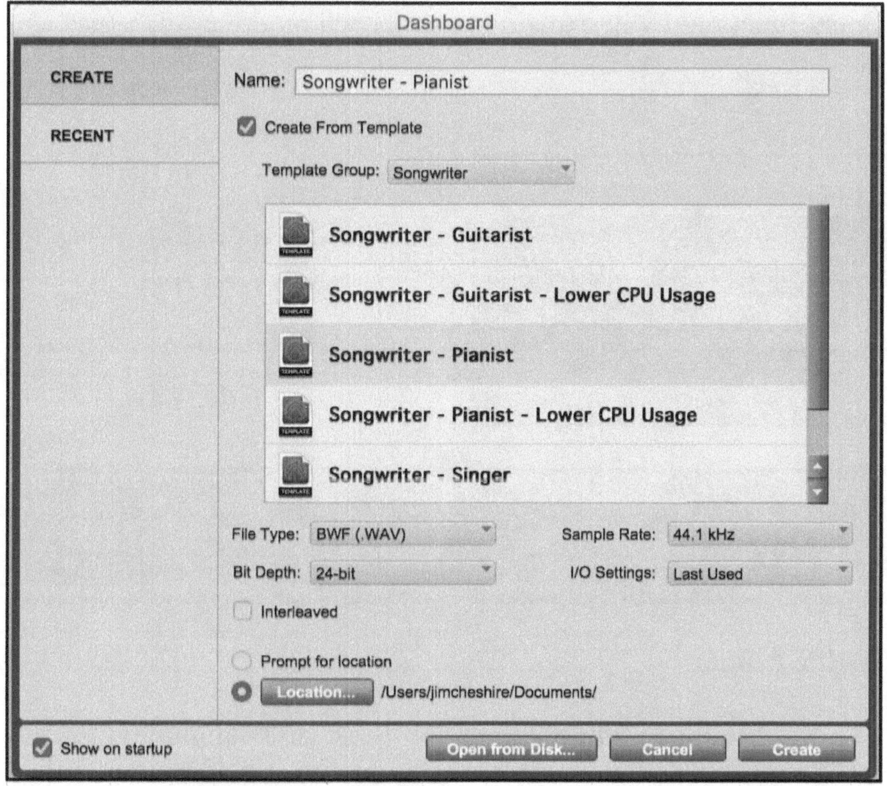

The dialog box will allow you to name and create a new session or pick and choose from a recently worked on session. It will also give you options to choose your file type, bit depth, and sample rate of the audio in which your session will create. After you name your session and decide on your settings, click create and your new, blank session will open.

Pro-Tools has four editing modes, editing in each mode provides a different function, I will explain in detail each mode:

1. Shuffle Mode (F1)

2. Slip Mode (F2)

3. Spot Mode (F3)

4. Grid (relative or Abso-lute) Mode (F4)

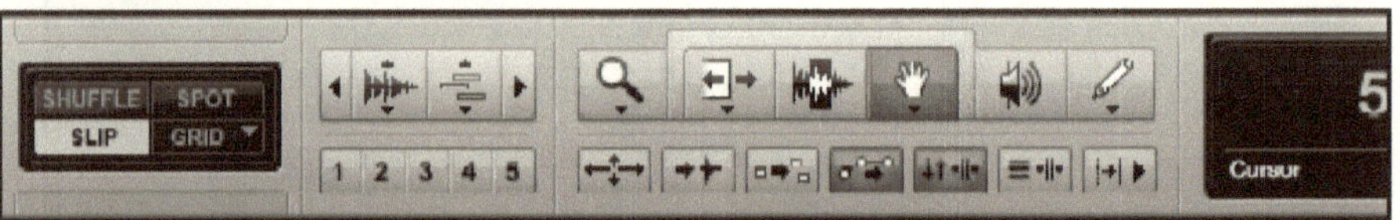

Shuffle Mode:

Shuffle mode will make region boundaries align next to each other, for example if you had two regions on a single track with a space between them, as soon as the right region was moved at all the start of that region would automatically snap to the end of the left region. The same process will apply when trimming regions as well, imagine three regions all directly following each other, if the end of the middle region was trimmed shorter, the region on the right would move in earlier to fill the space that was created. Usually described as a 'train-car' like snapping edit mode, this particular mode is very useful for ensuring that regions are placed directly next to each other without end-silence or overlapping occurring.

Slip Mode:

Slip mode will allow you to move and trim regions freely without affecting the position of any other regions within the session;

you can create overlaps or empty space between regions wherever you like. Unlike Shuffle mode, slip mode will allow you to trim region boundaries and only that region will be changed.

Spot Mode:

As you may be able to derive from its title, Spot mode allows you to precisely place or 'spot' regions to a very precise location value. As soon as you click a region in Spot mode you will be presented with a dialog box which asks for a precise value to place it within your session, you have the option to set the placement time of the regions start point, end point, or sync point. Instead of trying to nudge regions to their required location, this edit mode will allow you to place it with exact precision up to the finest time values. It should be noted however that edits such as trimming can be performed just as they would within Slip mode, the dialog only becomes present when setting its location placement.

Grid Mode (Absolute):

Within Grid mode, any region movements will snap to the nearest set time increment depending on which grid value is set within the edit window toolbar. It does not matter if you are moving the region, trimming region boundaries or drawing in MIDI information, it will all be snapped to the nearest time split value. We find this a particularly useful tool when changing a songs arrangement setup, for example a verse will usually last for a set number of bars, if the grid mode matches this value then we can easily grab the entire verse regions and move or copy them in bar values to later in the session without losing its time placement.

Grid Mode (Relative):

As mentioned briefly earlier, Grid mode will allow you to work in both Absolute and Relative mode so we had best explain the differences... When a region is moved or trimmed in Absolute Grid Mode it will snap to the nearest 'absolute' grid increment, the easiest way of explaining an absolute position is saying that the very beginning of the session will be the Start of the grid and the increments go up from there. Relative Grid Mode is used when we want to move a region which is not snapped to the absolute grid, for example a region which is perhaps placed a few milliseconds before a grid increment, with Relative mode enabled we can move that region by the set grid value while still maintaining the regions start/end position in relation to the absolute grid.

Now that we've discusses the four editing modes, to the right of the mode selections there are 6 tool selectors which represents:

1. **Zoom or Magnifier Tool (F5)**
2. **Trim Tool (F6)**
3. **Selector Tool (F7)**
4. **Grabber Tool (F8)**
5. **Scrubber Tool (F9)**
6. **Pencil Tool (F10)**

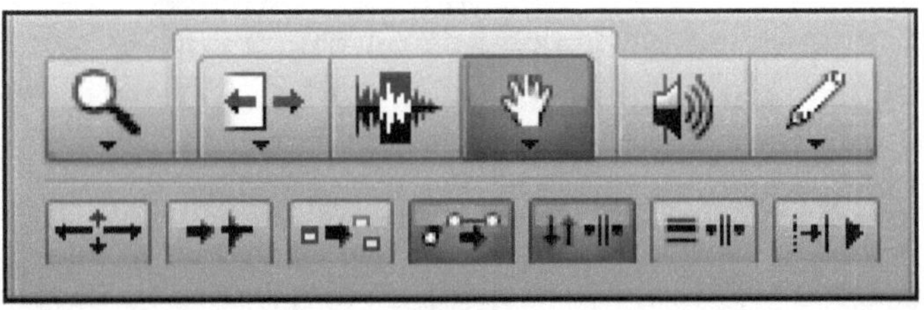

Zoom Tool: Allows you to click and drag on a certain area on your edit window to zoom in as precise as a sample or tick.

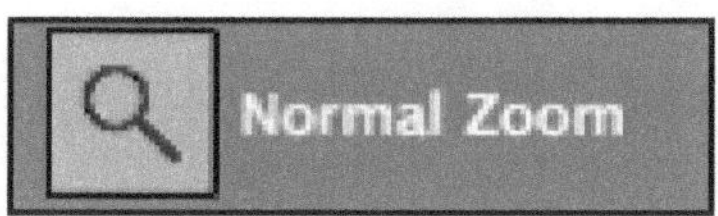

Normal mode: The zoomer tool remains selected after it's used.

Single zoom mode: The previously selected tool is automatically reselected after the zoomer is used.

Trim Tool: Allows you to trim or extend a audio, midi, or video region by clicking on the region on either side and dragging it either to the left or right. Click and hold the tool to access time compression/expansion, trim loop, or trim scrub tool.

Trim time compression / expansion tool: Lets you time compress and expand regions by clicking and dragging the region.

Trim scrub tool: Will allow you to scrub while you make your trim selection. Once you release the mouse, you will initiate the trim.

Trim loop: will loop regions when you click and drag the edges of a region

Selector Tool: Allows you to select your playback position or to highlight regions horizontally and vertically by clicking and dragging to make edits such as cut, copy, and delete. You can select multiple tracks at once

Grabber Tool: Allows you to grab a region and move it anywhere though out your session. If you double click on a region, it will allow you to also name the region.

Separation Grabber Tool: Allows you cut and paste an edit selection leaving a gap in the original region. First use the selector tool to make the selection then use the separation grabber to move the selection.

Object Grabber Tool: Allows for the selection of non contiguous regions

Scrubber Tool: Allows you to hear any waveform or region simply by clicking and dragging the scrubber over the region or waveform

Pencil Tool: Allows you to redraw a waveform for correction. You can also edit midi data and automation with this tool.

Now that we have learned the four edit modes and the six tools and a few shortcuts, before we begin to record, edit, and playback audio, let's discuss the transport. The transport is built into the edit window but can be opened in a separate window and moved around freely across your computer monitor by choosing the window tab in the menu and checking the selection "Transport". Your transport will contain your session information such as the tempo of your session, your pre and post-roll amount, and time signature. Your transport also contains buttons just like a tape machine to play, record, rewind, fast-forward, and stop. Let us get familiar with the transport as seen next.

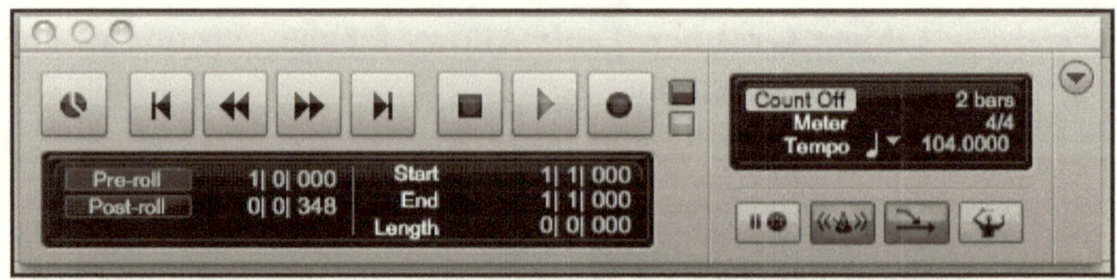

Now that you are familiar with the transport we can now create an audio track to record a voice over for testing purposes. There are two ways to create a track, the first is by selecting "Track" on the menu and select "new track" or you can use the shortcut (command + shift + N). A dialog box will open with track options, lets create a mono audio track and then hit "create" and a single mono audio track will appear in your edit window titles "Audio 1". You can rename it by double clicking on the track title and name it. See the example of the dialog box.

Make sure to select samples if you're recording vocals and select ticks if you are creating a MIDI or instrument track. We will discuss samples and ticks at the end of this chapter.

Now assuming that your DAW is set-up according to example #2 (in the signal flow chapter), your microphone's input should default to input #1. Click on your track view option box as seen below and make sure to show you "I/O" which stands for INs and OUTs.

Assign your input (IN) to the input of the microphone, which should be input #1. If you click on the record button within the track controls, just under the track name, the button should glow red, which indicates that you are ready to record. You should also notice your meters moving since your microphone is live. If your meters are not moving, you can do some simple troubleshooting by turning on the +48v phantom power if you are using a condenser microphone. Also check to make sure that your XLR Microphone cable is connected securely.

Below are three easy steps to set-up your track for recording audio. Within this option box, you have the option to view sends, inserts, I/O, and comments.

Step 1

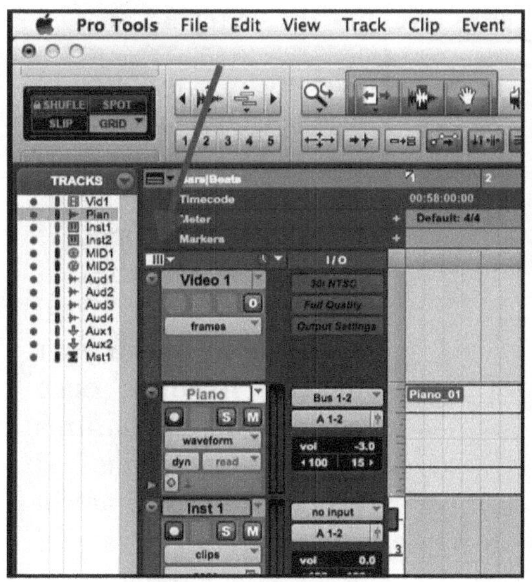

The arrow is pointing to a small option box. This bow will allow you to view your sends, Comment box, Inserts, Inputs and Outputs of each track. Click on this box and a dropdown menu will appear and select "I/O" to show your inputs and outputs.

Step 2

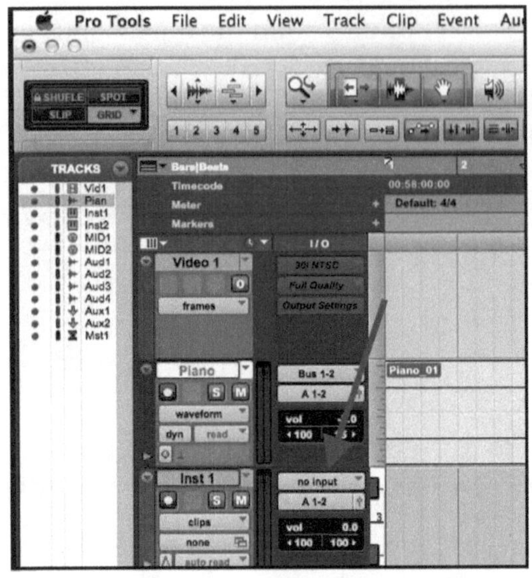

The arrow is pointing to the input selector for each track. Make sure to set the input for #1 since the microphone is connected to input #1 in the back of the ADC. If the input says "no input" please change it to input #1.

Step 3

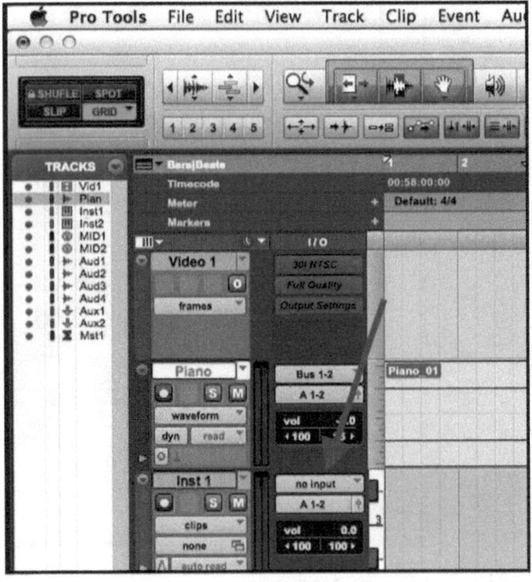

The arrow is pointing to a small circle, which is the record enable button. This button needs to be pressed in order to record, once enabled it will glow red. You should also notice your meters moving once enabled which confirms that you are connected properly and you are in fact receiving a signal from the microphone.

After completing these steps, you are now set up to record a test vocal through your microphone. The vocalist should be at least six inches away from the microphone and a pop filter should be used between the vocalist and the microphone. The pop filter will eliminate any accidental "pop" signals from the vocalist. Now that we have learned the basics of recording for Pro-Tools, let us discuss the technology behind digital recording and samples and ticks.

Capturing a sample of audio is like taking a picture. If you set up a camera to take a picture of the sky once every hour, you could only follow weather patterns in a very rough way. With this low "sample rate" you would probably miss many significant events. However, if you took pictures every second you would see much more detail (like the moment rain begins to fall). Digital recording is like taking pictures of music at a speed determined by the sample rate. If the sample rate in your session is 44.1 kHz, Pro Tools takes 44,100 pictures of your input audio every second. Each picture captures the amplitude (level) of the audio signal at that moment. Each sample is digitally mapped to an exact digital value and converted into binary digits (or bits). The number of bits in the system is referred to as bit depth. The higher the bit depth, the more accurate the digital representation of the analog sound. The precision of the amplitude value depends on the bit depth; small bit depths yield less precise representations of the audio signal.

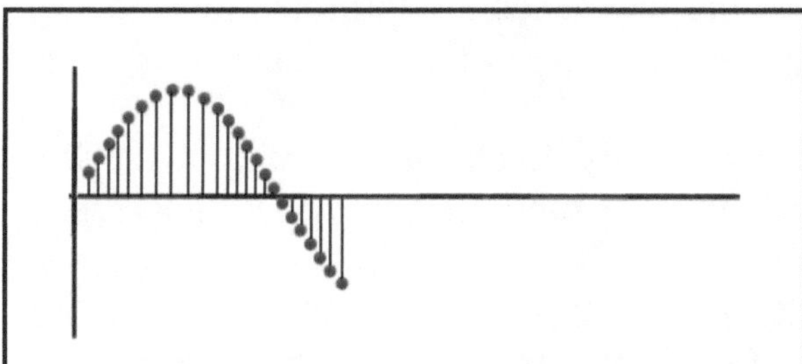

The dots represent samples recreating an analog sound wave and the lines are contains the bit information so just imagine the accuracy by placing 44,100 dots per second.

To use the photography analogy again, in color photography, a photo taken with a bit depth of 4 bits (2 to the 4th power) would only allow 16 different colors (2 to the 4th power = 2 x 2 x 2 x 2 = 16). If the color of a photographed object were not precisely one of the 16 colors allowed, the closest color would be assigned to it. Obviously, 16 colors can't possibly describe all of the shades and hues found in our colorful world.

The same logic applies to describing the myriad nuances of sound with an audio signal, usually, the more bits the better. That's why depths of 16 bits and 24 bits are offered in Pro Tools: 16-bit resolution offers 65,536 (or 2 to the 16th power) levels of audio amplitude. 24-bit resolution offers 16,777,216 (or 2 to the 24th power). With bit depths this high, it's like having thousands (or millions) of colors to choose from instead of only 16.

Now let us discuss what a tick is. Each quarter note in the Pro Tools tempo grid is divided into 960 subdivisions, called ticks. As a result, the duration of a tick varies according to the tempo of the session. If you're the mathematical type, here's the formula for determining a tick's length: One tick = 60,000 / tempo / 960 (where 60,000 refers to the number of milliseconds in a minute). If the tempo is 120 bpm, the duration of a tick is approximately 0.521 ms. faster tempos yield shorter tick values, while slower tempos have longer values. In comparison, the space between two audio samples recorded at a sample rate of 44.1 kHz is 0.023 milliseconds, a tick "slice" is longer than the distance between samples, so sample-based timing is more accurate in placing audio or MIDI events exactly where they were played (Of course, if you speed up the tempo of the session to approximately 2,756 bpm, then the ticks would be as short as the distance between

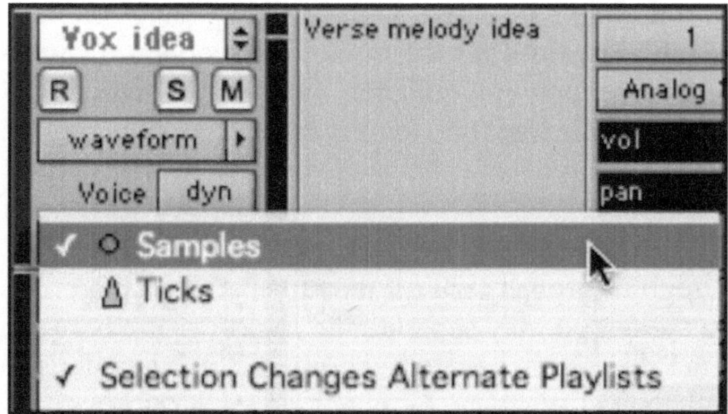

Samples and yield the same accuracy in representing audio digitally).

Sample and Tick Timebases with Pro-Tools, you can make any track either sample-based or tick-based. MIDI and Instrument tracks are tick-based by default, because MIDI events are usually locked to the tempo of the session. If the tempo of the session is changed, tick-based MIDI notes will follow the tempo changes and not lose their bar and beat location. Audio tracks are sample-based by default, and thus will not move if the session's tempo is changed. For example, a sample that is 3482950 samples away from the beginning of the session will always be the same distance from the first sample in the session's timeline, regardless of whether or not you change the tempo of the song.

Let's forget the default timebases for a minute. Let's make an audio track tick-based. The power of tick-based data is that audio and MIDI files can lock to the bars and beats of a session, even when the tempo of the session changes. As a result, making audio files tick-based can be quite a cool thing. Doing so lets you edit audio like you edit MIDI data. Making an audio track tick-based will fix the locations of your audio regions/loops to the Bar: Beats grid, and they will move relative to the sample timeline when tempo and meter changes occur. The audio region lengths will not change, just their starting points. Thus, the audio will move like the starting points of MIDI notes on tick-based MIDI tracks but will not change in duration like

MIDI notes do with respect to tempo and meter changes. Region groups that contain only one type of track in the group have the same timebase format (samples or ticks) as the tracks they contain. However, mixed-track region groups can include both sample-based and tick-based tracks. If you change the tempo of the session, region groups in tick-based tracks will adjust their length by moving all included regions accordingly.

Sample-based region groups will not move if the tempo is changed. Changing the tempo may separate the region group between sample-based and tick-based tracks. You can put all of the regions in a region group on the same timebase by changing each track's timebase or dragging the region group to a track with a different timebase. Changing the timebase creates a copy of the original region group. Both regions groups (the original and the copy) are listed in the Region List and have different timebases. When editing, if you notice that your audio regions/loops are drifting slightly out of time when repeated multiple times on a track, be sure to

Select and edit the region or loop with the Trimmer or Selector tool while in Grid mode (and using Bars: Beats as the Main Time Scale). Or type in exact locations in the Event Edit time locations. Either of these methods will ensure that the length of your region or loop is tight with the grid and not based on the length of the material in samples. Don't use the Grabber or double-click with the Selector. When the Main Time Scale is set to Bars: Beats, Pro Tools is tick-based, regardless of whether any track is sample-based. Consequently, some sample rounding will occur. You should be aware of this when you're creating/editing a loop and need the edit to fall right on the beat. When working with tick-based tracks, I recommend changing the Linearity Display Mode to "Linear Tick Display" instead of "Linear Sample Display (see below)." This toggle button is found next to the Ruler View Selector in the upper left side of the Edit window.

Toggle between Linear Tick and Linear Sample Display

Chapter 5

Editing in Pro Tools

Most of your Pro-Tools magic will happen in the editing window. To toggle between your edit and mix windows, use the shortcut (command & +) or simply by clicking on the window tab in your menu and select edit window. Below is an actual session's edit window.

1. This is your track bin. You can hide or show all created tracks within your Pro-Tools session by clicking on the track name and it will become hidden in your edit window but still remain in your session.

2. This is your edit mode selector, which we have discussed previously.

3. This is your tool section, when you click the bar above the 3 center tools (trim, selector, and grabber tools) you will notice that all 3 tools will turn blue which means that you are in "smart tool" mode. In this mode, the tool will change automatically, depending on where you place your mouse. Near an edge, it switches to the Trimmer tool. Close to the bottom edge, it becomes the Grabber tool. Close to the middle and top, it defaults to the Selector tool. In an upper corner, it changes to a Fade tool, click and drag to create an in or out fade. In a lower corner, it can create a cross-fade.

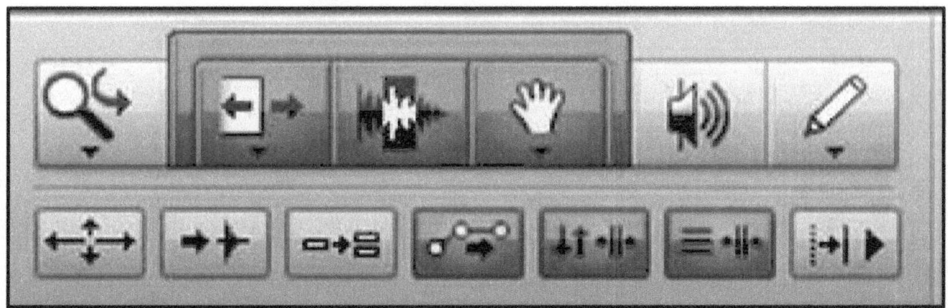

Smart Tool Mode

4. This is your counter window, which can be set to represent; minutes/seconds, Bars/beats, SMPTE Time code, samples, or Ticks.

5. This is your built-in transport, to separate your transport, select window from your menu then select transport.

6. This is your audio bin. Your audio bin will store all of your audio recorded within this session and can be opened by clicking the small arrow on the bottom of the bin's box. Audio can be dragged from this box and added directly to the timeline of your session. If you highlight an audio region within your timeline, it will also highlight the same audio file within the audio bin.

7. You will notice thin, colorful lines, which represents audio regions located within your session; this is referred to as your session's universe.

8. This is your comments window; you can move your cursor to the dialog box and type notes for each track as a reminder for something or mixing notes for the mix engineer.

9. These are your inserts, used to interrupt the signal's path by a plug-in. Your inserts section is for applying or routing an insert through your signal path of that channel or track. Refer to the signal flow charts for more information about inserts.

10. This is your sends section. This will allow you to route the signal from any channel to any other channel, usually an aux input for an effect. This routing send is done and transported to the aux destination by what is called a bus.

11. This is you I/O section, which contains your input, output, track volume, and track panning.

12. This is your timeline area where your waveforms will appear. Each track can be expanded or contracted by dragging the bottom of the track vertically further away or closer.

13. This is your Group area, you are able to group tracks by using the shortcut **(command & G)** then name and select which features that you'd like to group.

Now that we understand every area of the edit window, let us record some music or a vocal to a track. We will record enable the track and start recording by using these four different ways (command + spacebar) (decimal point on the numeric keypad) (F12) (Record + Play on the transport). Now that we have recorded a small test audio file, it should look like this below.

The displayed lines represent the recorded waveform, now we can edit this. Let us start by using the trim tool to the beginning of the waveform, which would be the left side. The image below will show the waveform after using the trim tool.

This image above is after we've taken the trim tool and clicked and dragged it a bit to the right, notice how the waveform to the far left has disappeared. Even though it does not appear on the track within the waveform but it is still present in the session within the audio bin, as described earlier.

An alternate method to getting rid of the first bit of waveform is by using the selector tool and clicking then dragging until you've highlighted the selection in which you want to delete. After you highlighted the section, press delete on your keyboard and it will disappear (following image).

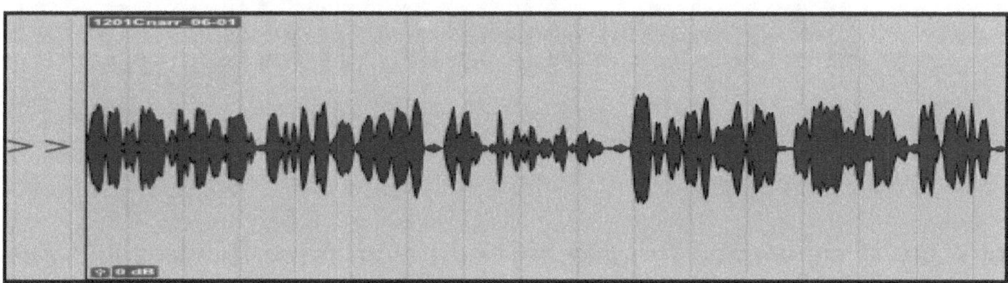

Now let us perform an edit in Shuffle mode and notice the difference in which the waveform moves as we delete the highlighted section. Below is the same waveform but instead of us highlighting the beginning of the waveform (Left side), we will select the middle section as seen below.

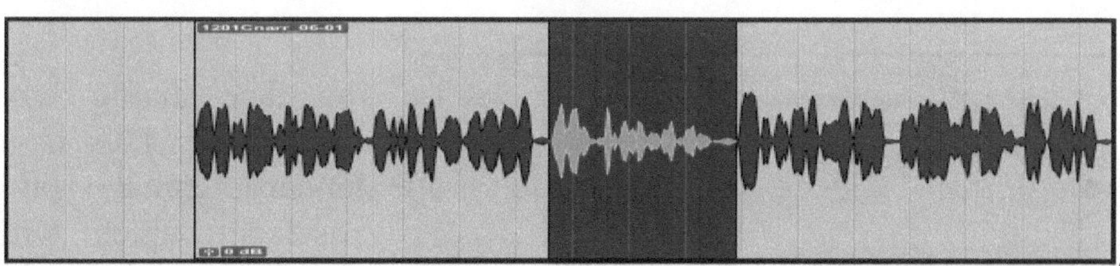

The next image will show us what happens when we delete the highlighted area in shuffle mode. It will delete the selected area and the left and right waveforms will adhere leaving a flawless gap as shown (following image).

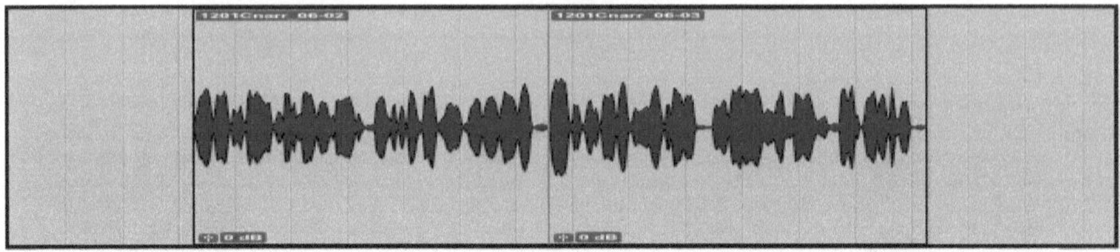

Let's say if we wanted to make this edit even more flawless and perform a crossfade at the adjoining regions, we would highlight the two regions as shown below. Notice how the center of our selection is somewhat directly in the center. This is where we want our crossfade to perform.

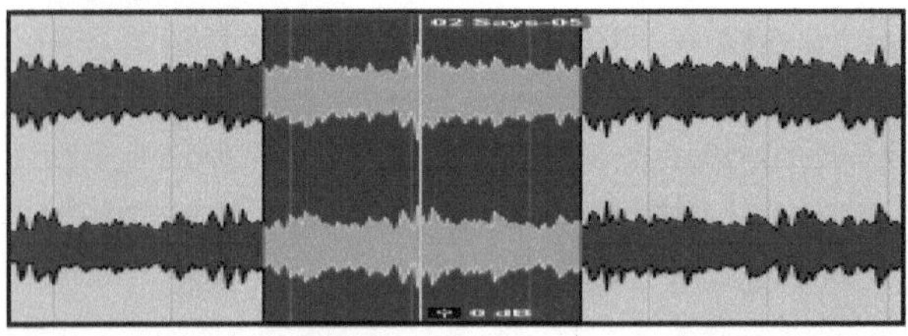

Highlight your crossfade selection, to perform a crossfade use the shortcut (command + F).

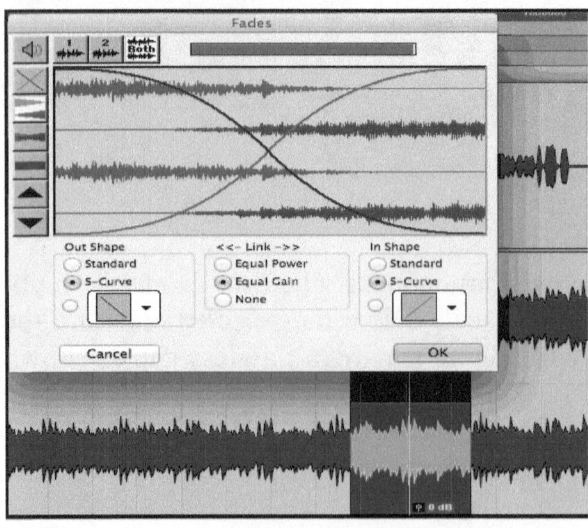

The face dialog box will appear. This is a standard equal gain crossfade, which will smooth out your edit even more to achieve a flawless sound. Hit the Ok button to complete the crossfade. Once the crossfade is done, it will resemble the next image.

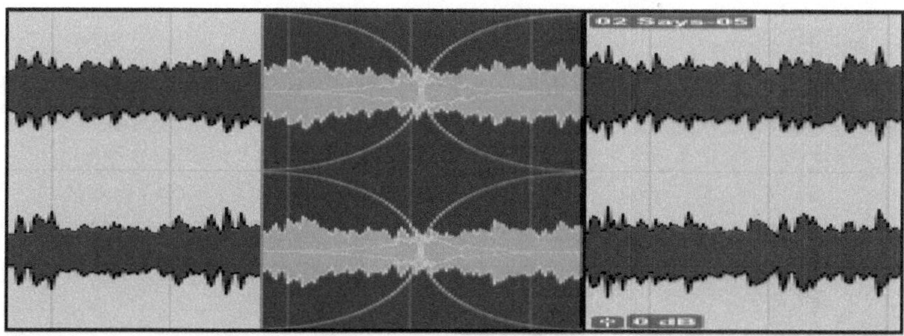

This is the waveform after the crossfade edit. This edit will make a flawless sound. Crossfading only works when two regions are adjacent to each other.

If a region is separated, by using the same shortcut (command + F) you will perform a regular fade as shown in the dialog box. You can either fade in or out, depending on where you highlight the audio region. You have options to use the standard shape, S-curve, or straight. The image shows the S-curve is selected on the audio will fade out as such.

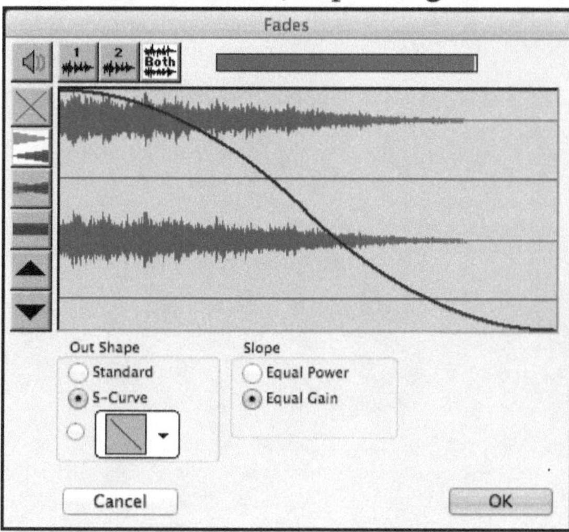

If you select a standard fade, the audio will fade out as described in the image below.

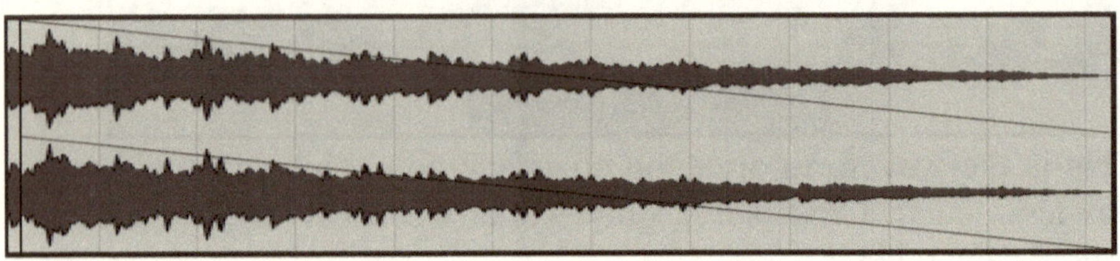

Chapter 6

Mixing in Pro Tools

Now that you've learned to navigate around Pro-Tools, record and edit, you should be rocking with multiple tracks, vocals, and instruments and your session look somewhat like this below but minus the sends and aux tracks. This is what we are going to learn now.

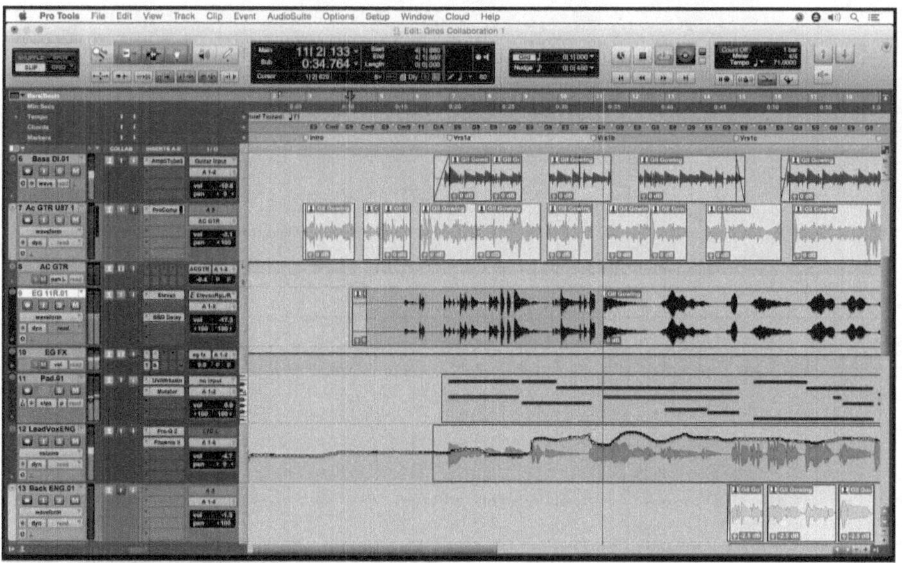

Let's say that we have 25 audio tracks, and they all need reverb in differing amounts. Surely it's just a question of instantiating a reverb plug-in on each track, which is not out of the question, but it'd be a terribly cumbersome way of working. First of all, the CPU hit of 25 reverbs could be colossal, potentially crippling if your Mac is already maxed out. Second, configuring all those reverbs is very time consuming and intensive, with way too much room for confusion. What's needed is a way to share one reverb flexibly amongst those 25 tracks, and it's easy. We're going to set it up on an aux track, which is a track type that you can't record on (although it can still be automated) and whose purpose in life is as a routing tool for receiving signals.

Next, we'll configure the specialized extra outputs on the audio tracks that are their aux sends, to split off some signal from each, destined for the reverb. We'll tie the whole lot together with a bus, an invisible audio conduit. Prior to explaining how this can be accomplished, I'd like to define a few key words which all work together.

- **Aux** – meaning auxiliary which you cannot record to but is the receiver track for your sends to output subgroups or effects such as reverbs, delays, flangers, Etc.

- **Send** – A send does exactly what it says. It sends the source signal from that original track to a receiving aux track.

- **Bus** – A bus is transportation or a method of internal routing or conduit, mostly used to send or bus a signal to an aux track.

- **Insert** – Routing the signal to a plug-in by interrupting the signal path

Below are the three steps to create your aux input tracks to set-up some effects and we will bus over some signals from a few sends.

1. Select track from the menu (command + shift + N) add a new track and create a stereo aux track. Set its output to your main hardware outputs (or whatever you use for monitoring), load your favorite reverb in one of its insert slots, and set the reverb's wet/dry mix to 100 percent wet.

2. Set the aux track input to an unused bus, let's choose bus 1-2.

3. For each of your audio tracks, click on the uppermost Send output pop-up and choose bus 1-2, which you'll find initially in the New Stereo Bundle submenu. Your send to you aux is now set-up correctly, now to put some reverb on an audio track, you just turn up its Send Volume knob (and optionally use its Send Pan knob too, to help with proper stereo imaging). Do this for any number of audio tracks, and bus 1-2 gathers all the sends together. Experimenting with different reverb types is so much easier now, because there's just the one plug-in to manage. As well as being more efficient, it also opens up new creative options. For example, it allows you to place an EQ in an insert slot below the reverb to really sculpt its sound. Try doing that for a reverb instantiated on an audio track and you'd end up equalizing the dry signal as well. You can also try compressing or gating the reverb output, or passing it through a chorus plug-in for a dreamy, de-correlated sound. If you toggle between you edit window and mix window, this can be done in the window option on your menu and click mix or you can simply use the shortcut (command and +).

After you set up your Aux tracks and sends, your mix window should look like this below.

Now repeat the steps from one and create a few more aux input tracks and a few more send via more unused busses. Add different effects on the inserts of your aux tracks, which is the beginning process to mixing. During the mixing stage, you will more than likely use a ton of dynamic plug-ins such as compressors and equalizers, make sure to send the vocal tracks that Should have the same EQ sound to the correct aux input tracks for example; after you blend all of your backing vocals, you will bus all of them to the correct aux input 1 and 2. Let's say your lead vocal should have a different sound so you will bus it to a different aux input channel which would have different effects and dynamics across it's inserts. Sometimes you may want to use your aux input track for a group of instruments as opposed to a group of vocals. There is no different set-up at all, the protocol is the same and below is an example of a drum sub-group (sending all of the drums via bus to an aux input). This is done for many reasons, maybe you want to control the levels of all of the drums or maybe you'd like to add an effect or mute all of the drums at certain spots throughout the composition.

Chapter 7

Import/Export Audio Files

Let's imagine that you have an instrumental version to your favorite song and would like to record vocals to it. In order to complete this task, you would have to import audio into your session, there are two ways to import audio, you can click file on the Pro-Tools menu and click the option "Import" then click "Import Audio. If you choose to import your audio this way, a dialog box will appear as seen below. It will prompt you to find the audio file in which Intend on importing.

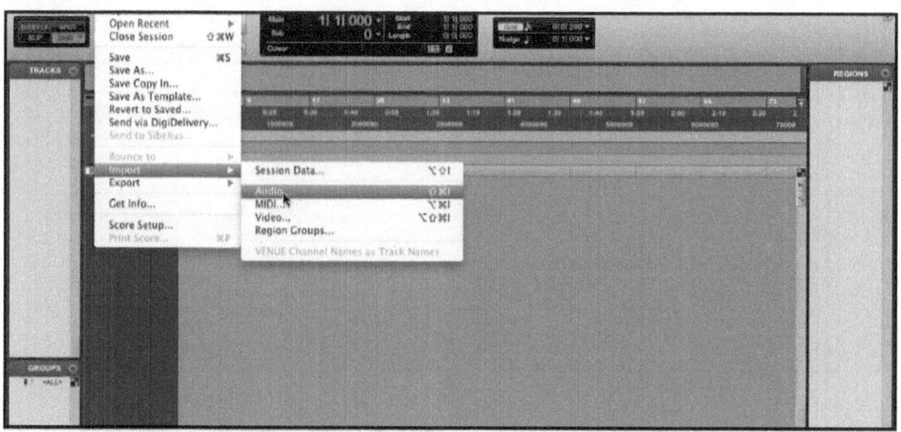

Notice within the dialog box, there is a play and a stop button located near the bottom of the dialog box, this is your audition control where you can play the audio file prior to importing it, making sure it's the correct file.

Files can be added, copied or converted for use in the session by using the "add" or the "copy/convert" buttons. The copy button will automatically change to "convert" if the selected file is not a type, which can be used natively within Pro Tools. Compressed audio files such as mp3s are not natively compatible and must be converted first before use. By clicking add, the audio file will remain in its original location and the session will reference that file at its original location (i.e. it is not in the project's audio files folder) by clicking copy, the audio file

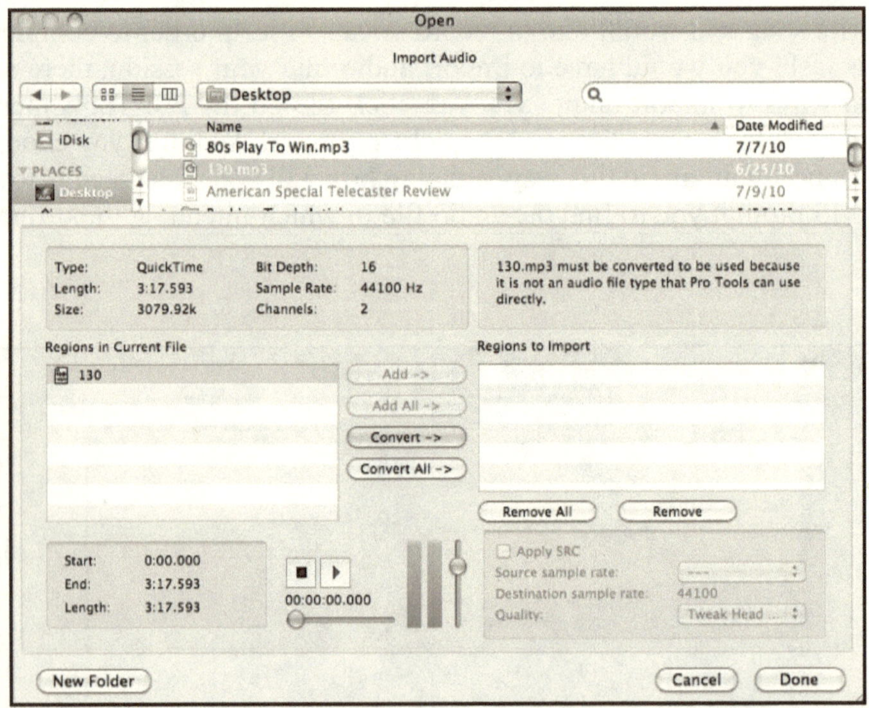

Is left in its original location and a copy is made in the project's audio files folder and it is this copy which is used by Pro Tools. By clicking convert, the audio file is left in its original location and a copy is made in WAV or AIFF format and this converted copy, which is stored in the project's audio files folder, is used by Pro-Tools.

If a file is compatible with Pro Tools you have a choice to add or to copy. Which is best? Well like most choices you are presented with it depends but for the overwhelming majority of new users I would recommend always copying.

In a Pro Tools session you can freely combine files with different bit depths but files with a different sample rate to the session would play back at the wrong pitch if they were imported directly. In this case a copy has to be made with appropriate sample rate conversion applied. In practice compatibility issues are easy to understand as the import window displays helpful information about the properties of the currently selected file and its compatibility with the session in the file properties section of the import audio window.

Next you would click the option "Convert or add" and notice, your file will appear in the bin to the right named "Regions to import". Finally click "done" and your audio will start to import. After the import conversion is complete, another dialog box will appear labeled "audio import option".

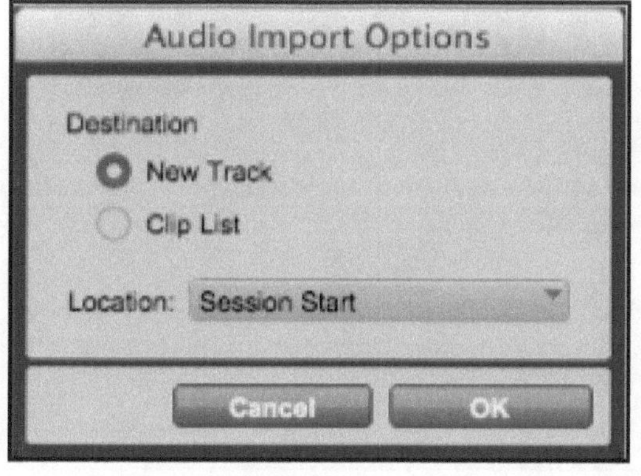

This box will ask if you'd like to add your imported audio into your session's timeline on a new track or would you like to add it into your audio bin for use later. Click "new track" and Pro-Tools will automatically create the track and you will see the region and waveform appear and the default track/region name will be the same as the file-name of the audio.

The second way to import audio is much easier "click and drag method" and convenient if you're already decided that you are using the import within your session. Minimize your edit window and locate the audio file in which you want to import, simply click and drag the audio file from it's source directly onto your timeline and you will notice the region box will appear and the waveforms will start to fill in as the audio is converted and imported.

Let's say that you have a fully completed song and would like to send it to someone or play it back in a MP3 player such as iTunes. You cannot open your pro-Tools session in iTunes and you will not be able to play or share your composition unless you export your audio files. Exporting tracks can be done as an individual audio file or you can export your entire session which is also called "bouncing to disk" either offline or real-time online. The first step is to use the selector tool and highlight the area on your timeline in which you'd like to bounce. Next you would click on "file" in the menu and then click on the option bounce to disk, as seen below.

Once you click "bounce to disk" option, another dialog box will appear which will contain your bounce options. First is your bounce source, which represents the source of your bounce for example; if your I/O main or master audio path has been assigned to outputs 3 -4 then you should select 3-4 as your bounce source. By default, your main and master outputs are assigned to outputs 1 and 2 so the dialog box is correct. Next are file type options, you can choose various file types such as MP3, AIFF, WAV, Etc. The next option is format meaning that your bounce can be exported as 2 files, left channel and right channel. To bounce both as a stereo 2-track, choose the option "Interleaved" or "Stereo Interleaved". Your next options are Bit depth

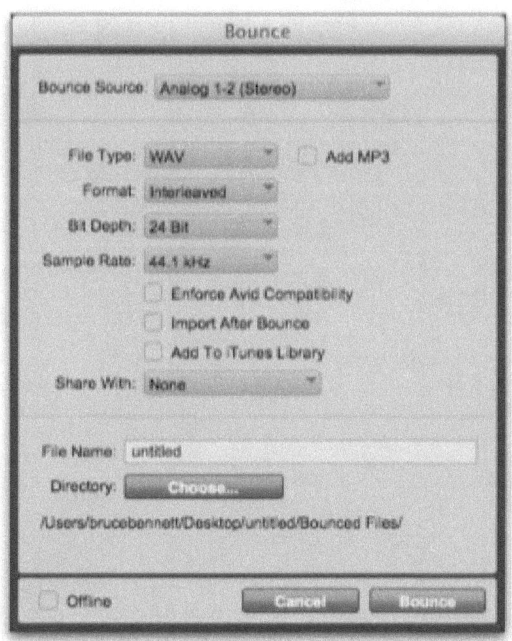

or Bit resolution and sample rate, which is self-explanatory. Next, name you file or track then choose which directory where you'd like your bounced file to, to make things easy, just choose your desktop. Notice carefully at the bottom left of the dialog box, there is a check box labeled "offline". By choosing offline, you will not hear your bounce in real-time and it will export as directed. I personally like to hear my bounces in real-time so I rarely bounce offline. Once you click the "bounce" button, another dialog box will appear notifying you that you are now bouncing as in the image. Once your file completes the bouncing stage, locate your file and play it back to make sure it's exactly as intended.

This is the bounce dialog box, which will appear while bouncing your session.

Chapter 8

Creating Music in Pro Tools

What makes Pro-Tools different from almost every other DAW is Digidesign (The original developers of Pro Tools) based the earlier versions of Pro-Tools around digital recording and editing and then later focused on the sequencing and MIDI aspects as opposed to the other DAWs how the focus was MIDI sequencing and later adding the features of recording and editing audio. Most music programmers prefer other DAWs to create such as FL studio or Logic Pro but ultimately your session would more than likely end up in Pro-Tools to edit and mix. Besides being the industry standard for editing and mixing, Pro-Tools has definitely improved their music programming functions and can eventually become one of the leading Music programming DAWs by top composers.

The first thing we will do is start by creating a mono Instrument track, which will be used for our metronome. Let's go to "Track" on the menu and select "new track" then you will see the above dialog box open.

Let's click the audio tab and change it to "Instrument Track" and the sample will automatically change to Ticks, which we've discussed in a previous chapter. Next, hit the create button and a new track will appear named "Instrument 1" represent that it is an instrument track and not an audio track. Make sure your inserts are viewable and click on an insert tab and select plug-in then select instrument and select click.

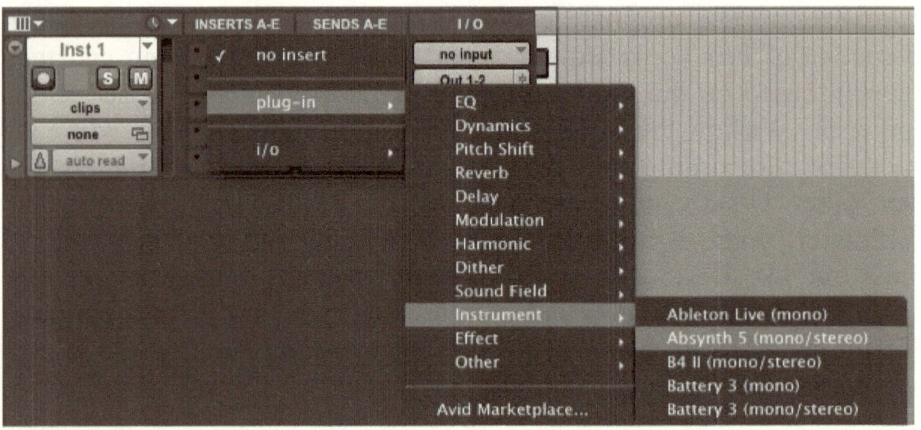

The click will open up and will look something like the image below, and the default tempo is always 120 BPM (beats per minute). The click also does a quarter-note pattern but you can change the parameters to eighth or sixteenth note as well as the volume of the accent click, which is usually the downbeat of the bar.

To change the tempo, you can perform the tempo change a few different ways but we will click Event then Temp operations and then constant, as pictured below.

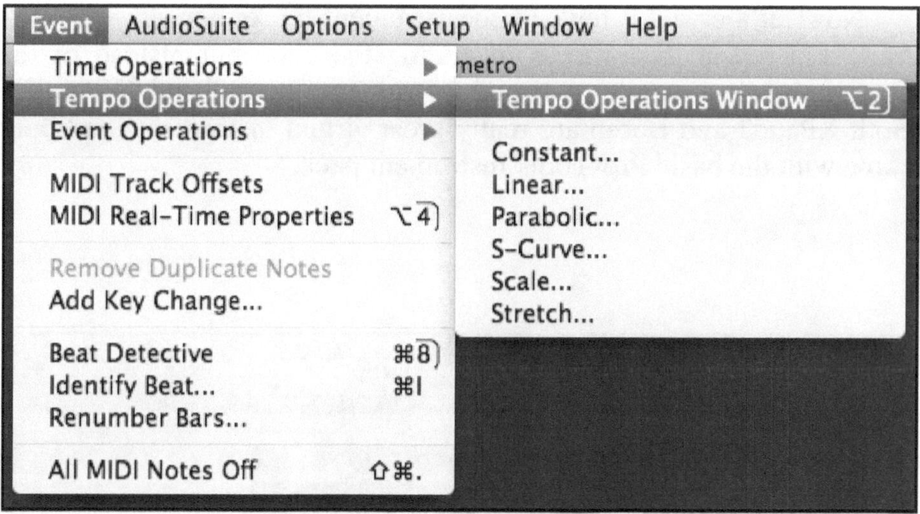

Once you click on the tab "constant" another dialog box will appear

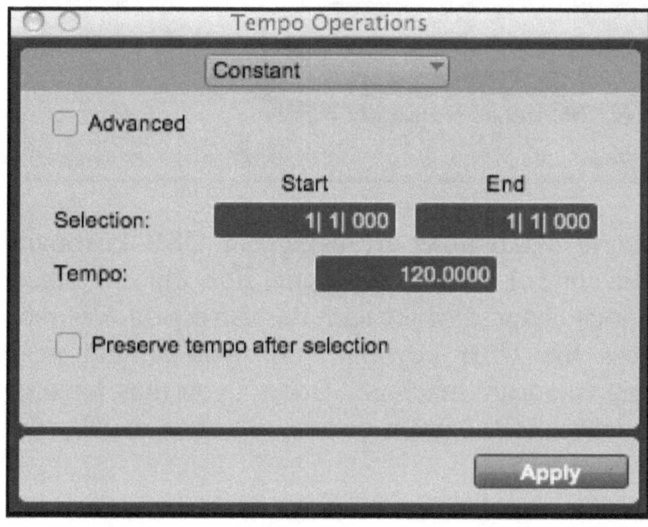

similar to the one on the left. If you want the tempo change to effect the entire session from bar 1 beat 1, make sure it looks similar to this. Highlight the tempo field and type in any desired tempos instead of 120 BPM then hit apply.

You will notice your metronome and click will adjust to your new tempo and if you have pre-recorded midi notes contained in your session, the tempo will also affect those notes.

Now let's create a few stereo instrument tracks beneath the click track and repeat the process of adding the click but instead of the click, let's add a couple of virtual Instrument "Boom" and "XPand2". Both XPand2 and Boom are really great virtual instruments and both come with the basic Pro-Tools Instrument pack.

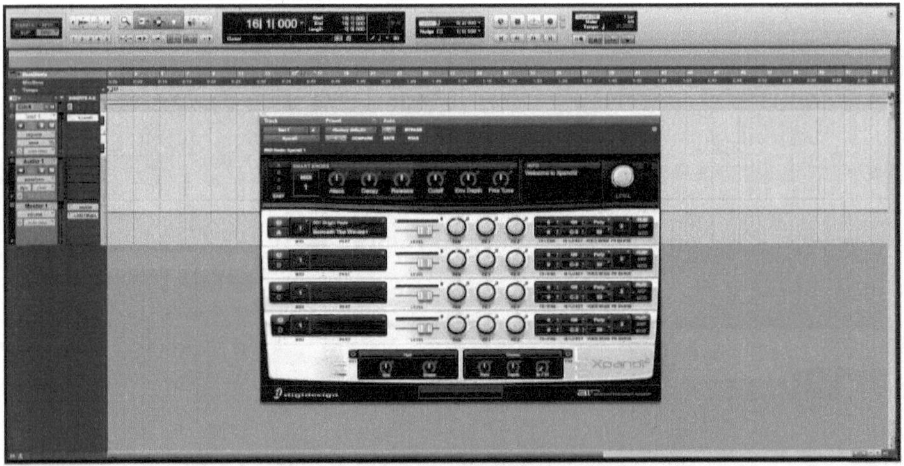

Let's record some simple MIDI notes by using our USB keyboard controller. To set-up the control, refer to my signal flow chart, method 3, which is in my previous chapter. Make sure the correct track is record enabled then press the USB keyboard and you should hear sounds. If you are using the drum machine "Boom" you may have to lower your controller octave to the drum octave, which is usually (-1 octave).

Once you start recording MIDI notes, you will not see waveforms as with audio signals but you will see small blocks as in the image below. These blocks or dots represent MIDI notes; which carries other information besides notes, it also carries volume or velocity data, from 0 to 127, sustain, expression, and any other midi information that may be recorded within those tracks.

Now that we have recorded two virtual instrument tracks, let's record more and your session will soon look similar to the one below. One of the best things about recording with MIDI Unlike audio is you can always change the sound because the MIDI notes are not sounds but information, which is then carried to the virtual instrument, which produces the sound. You are still able to use all of the 4 modes and tools as if we were editing audio waveforms.

As with and DAW, we can perform the same functions with Pro-Tools such as quantizing MIDI notes. To quantize notes, let's begin by selecting "Events" from the top menu then select Event operations and then click quantize then this dialog box will appear.

You can set your quantization up to 1/64th note as well as alter more quantization properties such as swing, triplets, and randomize.

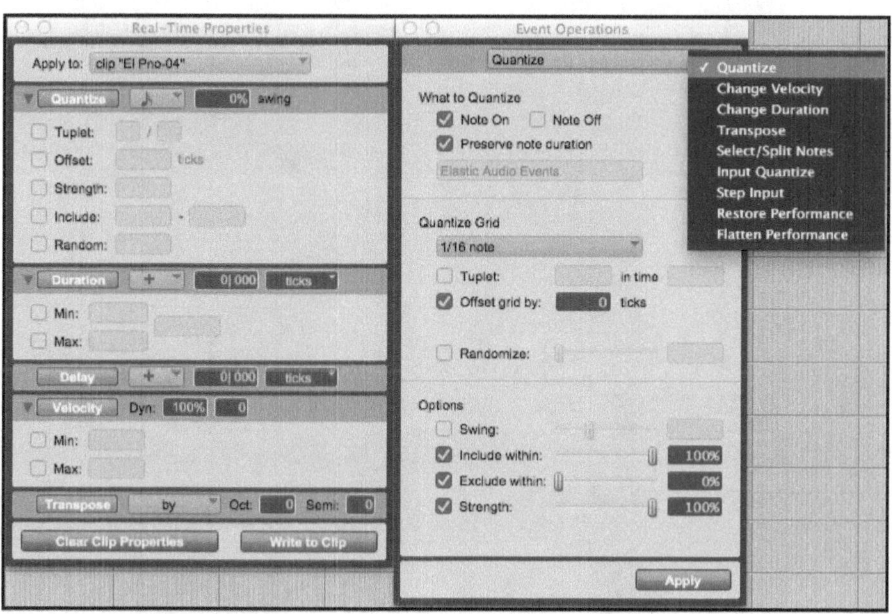

You can also change the view of your MIDI notes from notes to Velocity By clicking on your view To show velocity information as well as modulation wheel, sustain, mute, Etc. Select the edit pencil tool and you are now able to draw in your MIDI notes or MIDI information exactly like any other DAW. This feature is great for corrections or additions (see the following image).

Midi information can be edited with the pen tool

Chapter 9

Using Video Files With Pro Tools

I have used video directly with Pro-Tools for a music video where I had to insert sound effects after the artist decided to add CGI effects and moving objects. It was rather easy and I exported a new master music video with the synchronized sounds to match the visuals. You can also re synch new music to a pre existing video or do shorter or extended music video edits with the video capabilities. Pro-tools make it so easy as this version 12 is compatible with an array of different video files. When using video with Pro-Tools, make sure that your video is fully edited by your video editing software, as we will only work on the audio portion. There are three basic ways to work to picture in *Pro Tools*.

- First, you can lock *Pro Tools* via time code to an external video playback machine and then *Pro Tools* can 'chase' the VT machine. This process doesn't put any extra load on the computer but it is slow, because you have to wait for the VT machine to spool backwards and forwards and cue to the correct position on the tape before you can work on a section. You will also need some sort of SMPTE synchronizer like the Sync I/O.

- Second, you can import video files directly into a video track in *Pro Tools*. This is fast, because there's no waiting for the VT machine to catch up: *Pro Tools* can continue to work in a 'non-linear' way and will jump to the correct part

Of the video file as you move the cursor around the Session, just as it does with the audio files. However, handling the video file puts a load on the computer, and you will find that with high track counts and plug-in counts, *Pro Tools* is more sluggish when using a video file as your picture source.

- Third, you can use a separate non-linear video player. This can be either a second computer running an application or a dedicated non-linear player like Rosendahl's Bonsai Drive. This has the benefits of both of the first two options, with the only down side being cost.

I recommend importing your video files directly into Pro-Tools as it is so easy to work with and as simple as importing audio which we discussed in a previous chapter. Take a look at the images below, as I will explain step by step, the importing, editing, and exporting process while working with video.

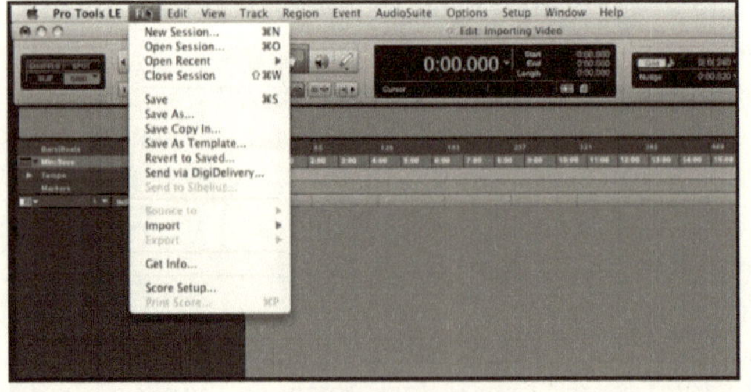

Click on File on your menu and scroll down to import, just as if we were importing audio.

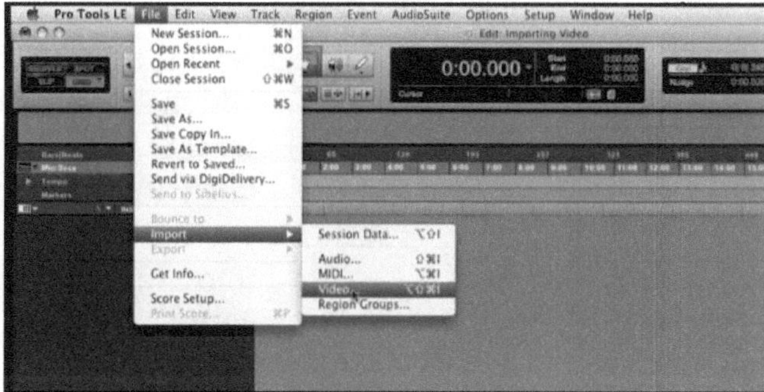

Next, click on video. Make sure to know or remember where your video file is.

After you click on video, it will prompt you to locate the video file you'd like to import. Once you find the file then click on "open".

Another dialog box will open and it will ask you where should you save the audio file, which is attached to the video. Just click choose which will add it into the default audio folder for the session.

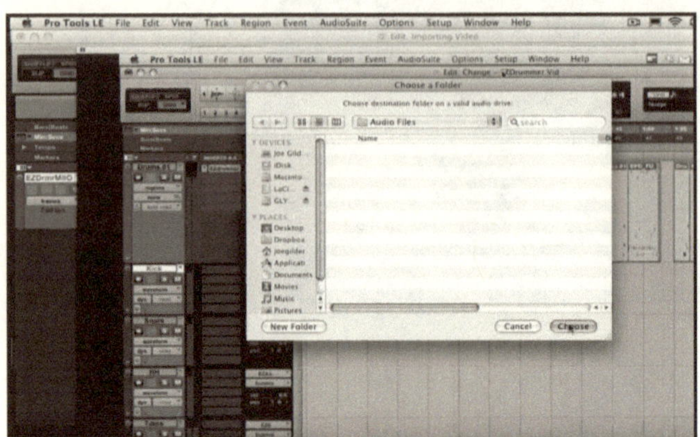

Next, a dialog box will pop up asking you for the start location of the video as well as importing the audio. It's always best to import the audio as a reference and then you can mute or delete it later.

Now you will see a dialog box showing the import progress. You will also see the video and audio waveforms develop as they import.

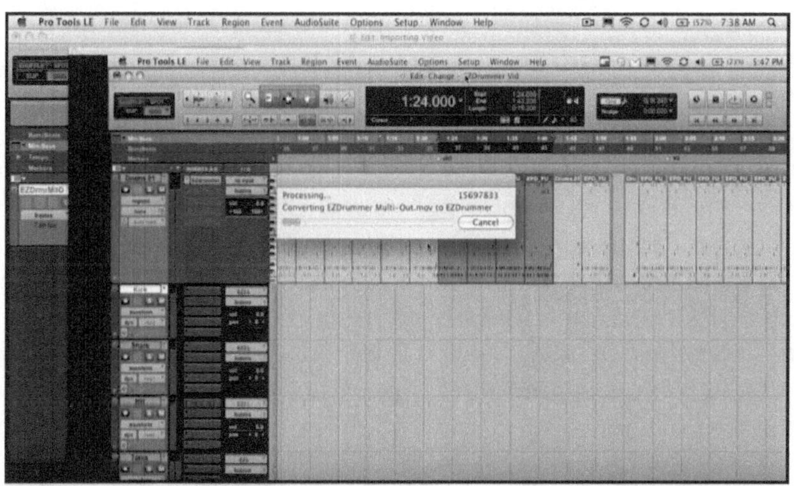

Once your video is imported, a video monitor will open and your video will play in that viewer, you will see video clips and the audio from the Imported video track will be directly beneath the video track.

You can now add effects, swap audio, shorten, or any type of edit you would normally do if you were just working on a regular Pro-Tools session or MIDI session. If there is already music contained in the video, you can perform overdubs and add more music or you can fully do a blown out post ADR session and still have access to all of the great features Pro Tools have to offer.

Once you're complete with your additions, edits, overdubs, or voice-overs, you'll have to export or bounce it just as you would with an audio session. Bouncing or exporting is pretty much the same but with video, there are a lot more options to choose from because there are so many different video formats and specifications. The first step in exporting your video track click "File" on your menu the choose bounce to Quick Time as seen in the image to your left. Don't forget as with the audio bounce, you must highlight your bounce area or your export may not be what you want.

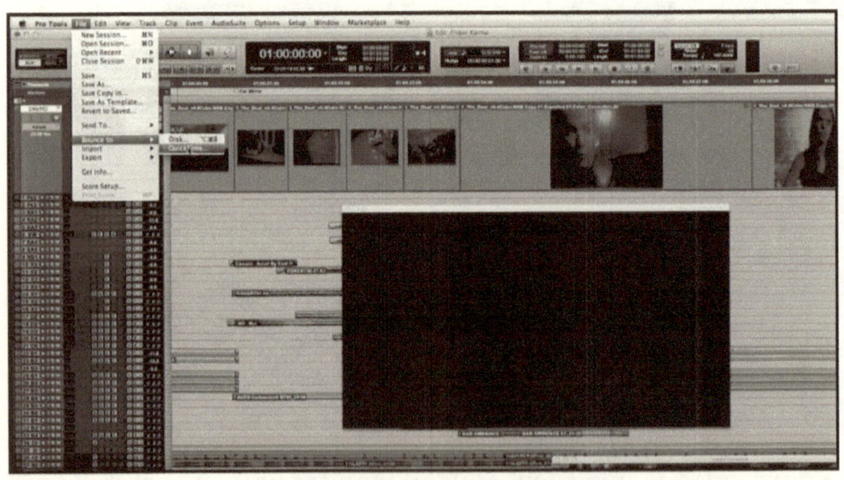

This dialog box will appear after you select bounce to Quick Time. You should be somewhat familiar with video formats and specifications. At this point, click quick time settings and another box will

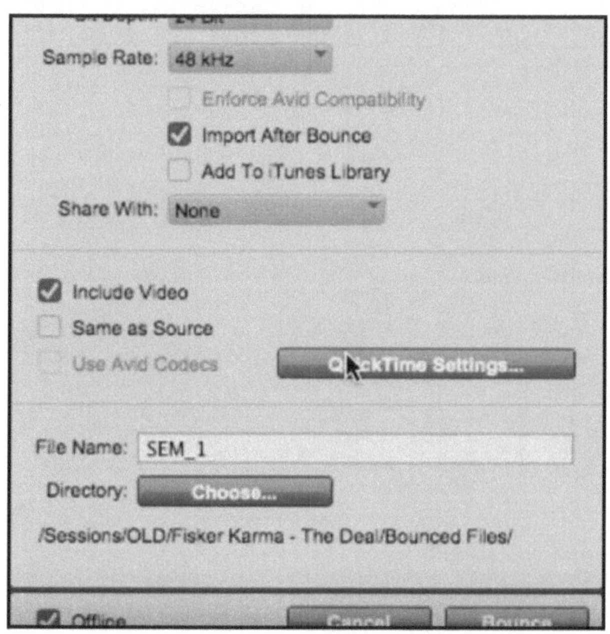

open where you can set the specifications of your video. The image (Left) is a close up of the dialog box for your audio settings. This is where you would name your bounce or export, set your bit depth, sample rate, directory, Etc. You also have the option to bounce offline as with the audio session from our previous chapter.

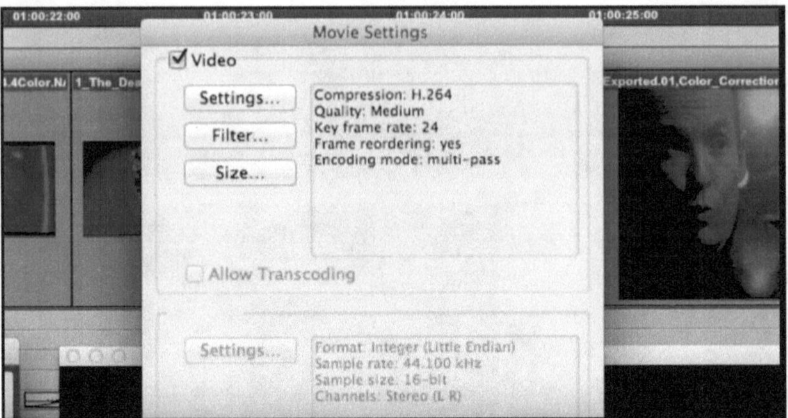

This is the final dialog box that will open after you choose quick time settings. You can set your compression, Filter, and size of your exported video. Once you are satisfied with your specifications then hit "Bounce" and your final video will begin exporting. Importing and exporting is just as simple with video as it is with audio and the quality is professional grade.

Chapter 10

Plug-Ins For Pro Tools

As a producer and Engineer, plug-ins are very essential and is my determining factor for my choice of DAW. This chapter will discuss how plug-ins work and the transformation of plug-in history.

Plug-ins are add-on software tools that provide additional signal processing functionality to your DAW. There's a plug-in for almost anything you can imagine: EQ, dynamics, reverb, sound synthesis, pitch and time shifting, noise reduction, mastering, surround encoding, and more. Some plug-ins emulates the sonic characteristics of hardware processors, amplifiers, or microphones. Others give you power and flexibility that is unique to the world of computer-based digital audio. To use plug-ins with Pro-Tools, you simply click on an insert button on any particular track, then search, find, and add. At that point, you're free to use presets, customize settings, and automate a given plug-in's activity within a Mix, any of those settings can be recalled at any time. Whether you're doing music production or sound design. The earlier Pro Tools plug-ins (prior to PT 10) were available in three formats: TDM, Real-Time AudioSuite (RTAS), and AudioSuite but Avid has now updated to AAX (Avid Audio Expansion) plug-ins, Native or DSP. This change was introduced with Pro Tools 10, making the TDM and RTAS obsolete.

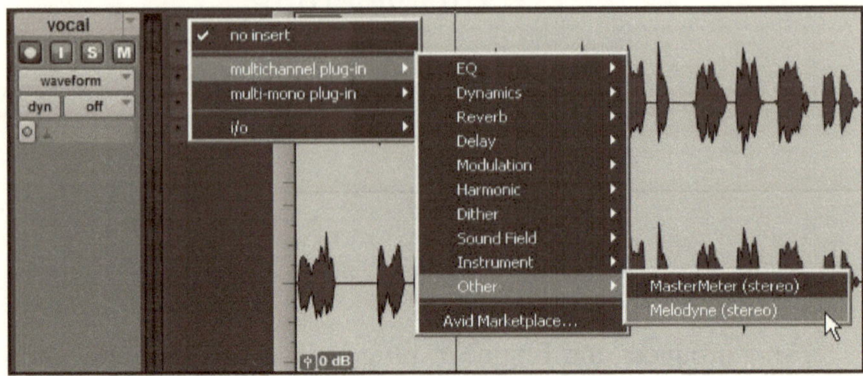

AAX
For Pro Tools version10 and later |HD systems

Avid Audio Extension Plug-ins has replaced both TDM and RTAS plug-ins since Pro Tools version 10. AAX plug-ins has become 64 bit and considered a hybrid because AAX functions similar to TDM and RTAS when referring to DSP and processing. AAX Native uses your host computer while AAX DSP uses dedicated cards for processing and is only compatible with HDX systems.

TDM
For Pro Tools version 10 and earlier |HD systems

TDM plug-ins are uniquely powerful because they use the dedicated hardware found on the DSP cards of Pro Tools|HD system rather than the limited resources of the host computer. As a result, TDM plug-ins are afforded much more processing power than what is available in any competing system. You can add TDM plug-ins up to the hardware DSP capacity of your system and get the same fast, reliable performance no matter how many tracks you have in a session. If you need more mixing horsepower, you can easily add DSP cards to your system.

RTAS
For Pro Tools|HD, Pro Tools LE, and Pro Tools M-Powered systems

Real-Time AudioSuite (RTAS) plug-ins are host-based, relying on the host computer's processing power to process audio in real time. Functionally, RTAS plug-ins offer many of the same benefits of TDM plug-ins. Their parameters can change in real-time, they can be fully Automated, and their effects are not permanently written to the audio file. Since they are host-Based, RTAS plug-ins shares computer-processing resources with a number of other functions. As a result, the number and variety of RTAS plug-ins you can use is dependent on other host-Dependent factors, such as track count (on Pro Tools LE systems), edit density, and the amount of automation in a session.

AudioSuite
For Pro Tools|HD, Pro Tools LE, and Pro Tools M-Powered systems

AudioSuite plug-ins employ file-based processing. Depending on how you configure a non-real-time AudioSuite plug-in, it will either create an entirely new audio file or alter the original source audio file with the processing included. AudioSuite plug-ins are useful for conserving DSP power. They are also suitable for certain types of processing where there is no real-time benefit or application, such as normalization or time compression.

VST
For Cubase, Nuendo, and Wavelab

Back in 1996, Steinberg, the German software developer behind hugely successful music programs like Cubase, Nuendo, and Wavelab, developed a technology called VST. VST stands for Virtual Studio Technology. The first thing Steinberg did with this technology was to incorporate it into their flagship music software, Cubase VST. All of us who are producing Music in our home studios, in one-way or another, owe respect to Steinberg. Cubase VST was the very first music application that offered real-time host-based plug-in processing. It offered a then powerful and affordable way to produce music "in the box", using only the software and a decent computer. At the time when VST came out, Digidesign Pro Tools did not have a host-based, real-time plug-in format of their own. So, if you wanted to Run Pro Tools with real-time plug-ins, you would have to use Pro Tools TDM, with its proprietary hardware DSP cards. Unfortunately for home studio enthusiasts, the typical Pro Tools TDM system in 1996 cost between $10k and $20k.

In September 1999, Digidesign announced their own host-based real-time plug-in format called RTAS. With RTAS, Pro Tools users could now have real-time, host-based plug-in processing in an affordable Pro Tools home version, Pro Tools LE. There are tons of amazing VST plug-ins and Steinberg made the Software Development Kit for VST freely available to anyone who wants it. Therefore, any experienced developer can create a processor or virtual instrument and make it into a VST plug-in. As a result, there are thousands of VST plug-ins out there, including free, low-cost, and/or commercially available ones. The sheer quantity of VST developers and plug-ins out there means that finding great VST plug-ins for free requires some time and effort. But in spite of that, it's worth it.

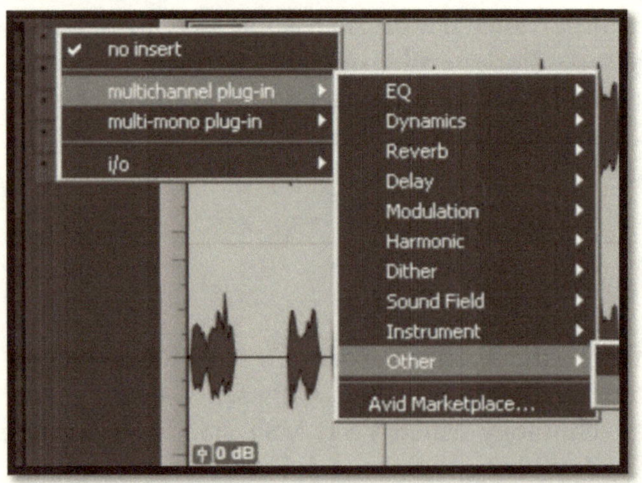

When you click to insert a plug-in, you will notice a Multi-channel plug-in selection and a Multi-Mono plug-in selection. Each selection will have an array of categories and within each category you will find your list of individual plug-ins.

Pro Tools Plug-in categories

- **EQ**
- **Dynamics**
- **Reverb**
- **Delay**
- **Modulation**
- **Harmonic**
- **Dither**
- **Sound Field**
- **Instrument**
- **Other**

Prior to us moving on, understand the difference between Multi-channel plug-in selection and Multi-Mono plug-in selections. When using Multi-mono, this will allow you to edit the parameters of a stereo plug-in (left and right) independently of each other, which can add a cool effect.

EQ (Equalizer) – There are numerous types of EQ's and they are used to adjust frequencies to enhance or sculpt an audio signal. There are 3 band EQ's, 4, 5, and 6 bands as well. Each plug-in has it's own analog replication and technology and engineers are tuned into their specific brands and models of choice.

Dynamics – This category is for altering the sound dynamics such as compression/Limiter, Expander/Gates, and DeEssers. You can control every aspect of dynamic levels such as attack, release, hold, Threshold, and ratio.

Reverb - a reverb is created when a sound or signal is reflected causing a large number of reflections to build up and then decay as the sound is absorbed by the surfaces of objects.

Reverb is the most common audio effect, partly because it is used in so many situations from music studios to television production. Every sound operator should have a good understanding of reverb and how/when to apply it. There are multiple types of reverb plug-ins as well as reverb settings to replicate the reflecting surfaces.

Delay – A delay is basically an echo but not as simple as an echo if you understand. You can set multiple parameters within each delay plug-in to fill empty spaces in music or vocal tracks.

Modulation - Modulation effects have been used intensively for decades: voice doubling, sound widening, chorus effect, flangers, warble effects, and vibrato. They are still heavily used in many different styles of music to process voices, guitars, and synths or even drum beats.

Harmonic – These plug-in effects will alter anything pitch related such as Lo-Fi, Air Distortion, Air Enhancer, Etc.

Dither – These are rarely used plug-ins but Dither is a form of randomized noise, which when added to a digital recording masks the less preferable quantization noise created by reducing bit depth. Such as the dither or POWr Dither plug-ins.

Sound Field – These plug-ins are also rarely used such as Air Stereo Width, Down Mixer, SurroundScope and PhaseScope

Instruments – This category will contain all of your virtual instruments for sequencing MIDI such as Boom, Xpander2, Piano, Etc.

The following are some popular Plug-Ins which are most used in to-day

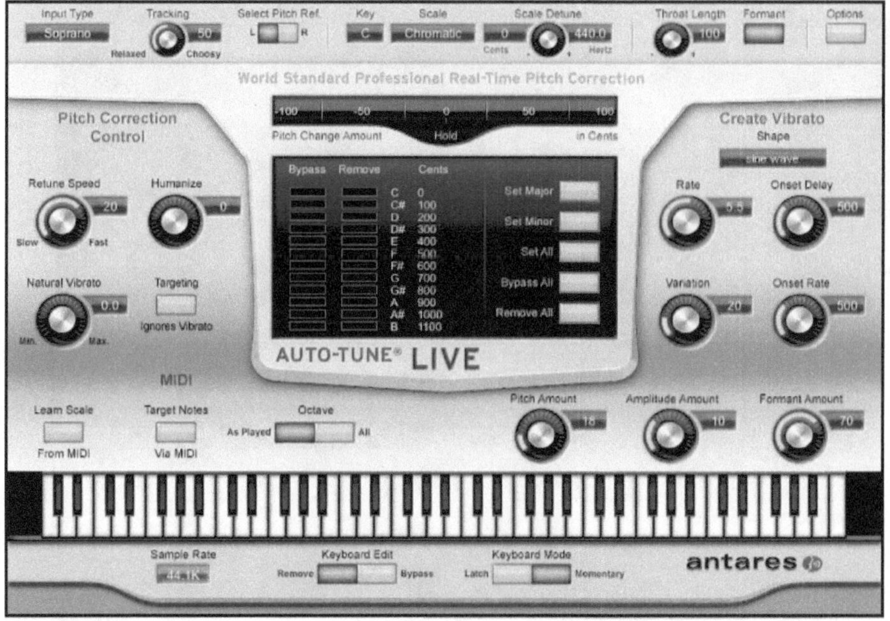

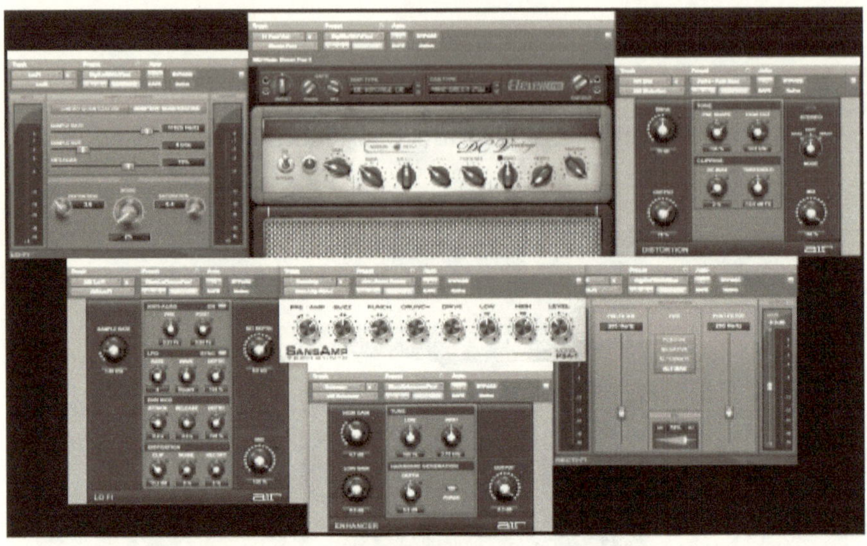

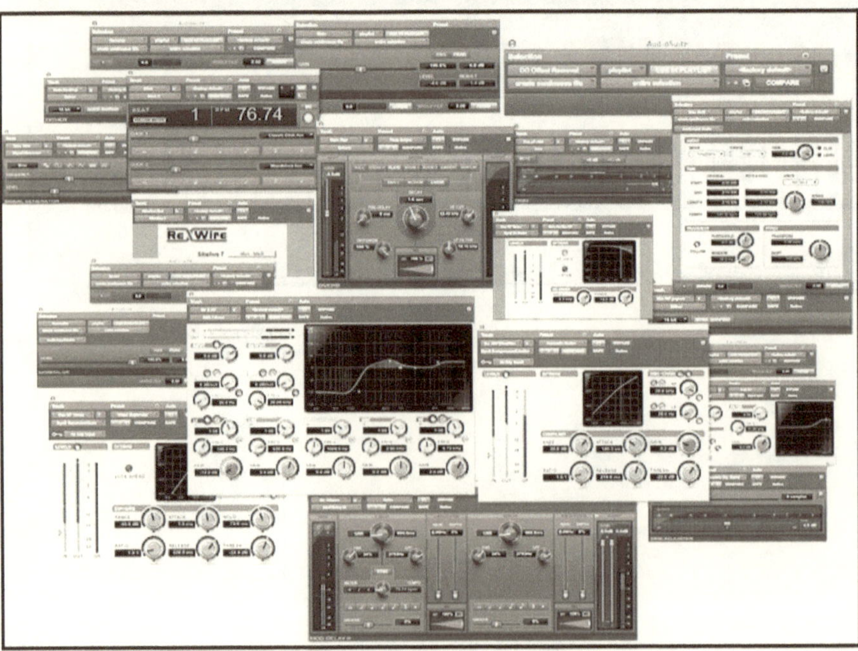

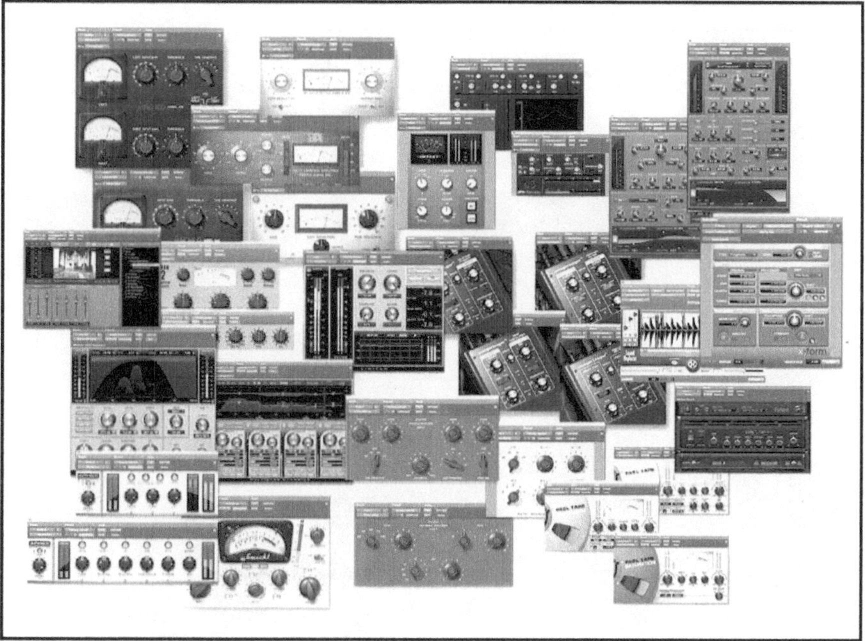

Chapter 11

Shortcuts For Pro Tools

1. File Menu	
Ctrl+N	New session...
Ctrl+O	Open session...
Ctrl+Shift+O	Open recent
Ctrl+Shift+W	Close session
Ctrl+S	Save
Ctrl+Alt+B	Bounce to - Disk...
Alt+Shift+I	Import - Session data...
Ctrl+Shift+I	Import - Audio...
Ctrl+Alt+I	Import - MIDI...
Ctrl+Alt+Shift+I	Import - Video...
Ctrl+P	Print score...
Ctrl+Q	Exit

2. Edit Menu

Ctrl+Z	Undo
Ctrl+Shift+Z	Redo
Ctrl+Alt+Z	Restore last selection
Ctrl+X	Cut
Ctrl+C	Copy
Ctrl+V	Paste
Ctrl+B	Clear
Start+Shift+X	Cut special - Cut clip gain
Start+Shift+C	Copy special - Copy clip gain
Alt+M	Paste special - Merge
Ctrl+Alt+V	Paste special - Repeat to fill selection
Ctrl+Start+V	Paste special - To current automation type
Start+Shift+B	Clear special - Clear clip gain
Ctrl+A	Select all
Alt+Shift+5	Selection - Change timeline to match edit
Alt+Shift+6	Selection - Change edit to match timeline
Alt+[	Selection - Play edit
Alt+]	Selection - Play time-

	line
Ctrl+D	Duplicate
Alt+R	Re-peat...
Alt+H	Shift...
Ctrl+Shift+E	Insert silence
Ctrl+T	Trim clip - To selection
Alt+Shift+7	Trim clip - Start to selection
Alt+Shift+8	Trim clip - End to selection
Ctrl+E	Separate clip - At selection
Ctrl+H	Heal separation
Alt+Shift+3	Consolidate clip
Ctrl+M	Mute clips
Ctrl+U	Strip silence
Alt+Shift+U	TCE edit to timeline selection
Ctrl+Alt+H	Automation - Copy to send...
Ctrl+Alt+T	Automation - Thin
Ctrl+/	Automation - Write to current
Ctrl+Alt+/	Automation - Write to all enabled

Ctrl+Shift+/	Automation - Trim to current
Ctrl+Alt+Shift+/	Automation - Trim to all enabled
Alt+/	Automation - Glide to current
Alt+Shift+/	Automation - Glide to all enabled
Ctrl+F	Fades - Create...
Alt+D	Fades - Fade to start
Alt+G	Fades - Fade to end

3. View Menu

Ctrl+Alt+M	Narrow mix

4. Track Menu

Ctrl+Shift+N	New...
Ctrl+G	Group...
Alt+Shift+D	Duplicate...
Alt+K	Set record tracks to input only
Ctrl+Alt+F	Scroll to track...
Alt+C	Clear all clip indicators

5. Clip Menu	
Ctrl+L	Edit lock/unlock
Alt+Start+L	Time lock/unlock
Alt+Shift+B	Send to back
Alt+Shift+F	Bring to front
Ctrl+Alt+Start+NumPad0	Rating - None
Ctrl+Alt+Start+NumPad1..5	Rating - 1..5
Ctrl+Alt+G	Group
Ctrl+Alt+U	Ungroup
Ctrl+Alt+R	Regroup
Ctrl+Alt+L	Loop...
Ctrl+R	Capture...
Ctrl+Alt+Shift+R	Rename...
Ctrl+,	Identify/Remove sync point
Ctrl+0 (zero)	Quantize to grid
Alt+NumPad5	Elastic properties

6. Event Menu	
Alt+NumPad1	Time operations - Time operations window

Alt+NumPad2	Tempo operations - Tempo operations window
Alt+NumPad3	Event operations - Event operations window
Alt+0 (zero)	Event operations - Quantize...
Alt+P	Event operations - Change duration...
Alt+T	Event operations - Transpose...
Alt+Y	Event operations - Select/split notes...
Alt+NumPad4	MIDI real-time properties
Ctrl+NumPad8	Beat detective
Ctrl+I	Identify beat...
Ctrl+Shift+.	All MIDI notes off

7. Options Menu

Alt+L	Loop record
Ctrl+Shift+P	QuickPunch
Ctrl+Shift+T	TrackPunch
Ctrl+J	Transport online
Ctrl+Shift+J	Video track online
Ctrl+K	Pre/post-roll

Ctrl+Shift+L	Loop playback
Ctrl+Start+P	Dynamic transport
Shift+/	Link timeline and edit selection
Ctrl+Alt+P	Auto-spot clips
Start+Shift+T	Edit/Tool mode keyboard lock

8. Setup Menu

Ctrl+NumPad2	Session

9. Window Menu

Ctrl+Alt+J	Configurations - Window configuration list
Ctrl+Alt+Start+W	Hide all floating windows
Ctrl+W	Close window
Ctrl+=	Mix
Start+=	MIDI editor
Alt+Start+=	Score editor
Alt+=	MIDI event list
Alt+N	MIDI editors - Bring to front
Alt+Shift+N	MIDI editors - Send to back
Alt+'	Task manager

Alt+;	Workspace
Alt+O	Project
Alt+J	Browsers - Bring to front
Alt+Shift+J	Browsers - Send to back
Ctrl+NumPad1	Transport
Ctrl+NumPad3	Big counter
Ctrl+NumPad4	Automation
Ctrl+NumPad5	Memory locations
Ctrl+NumPad7	Video universe
Ctrl+NumPad9	Video

10. Misc.

Shift+Space	Half-speed playback
Ctrl+Shift+Space, Shift+F12	Half-speed record

11. Numeric Keypad Shortcuts

NumPad0	Play/Stop
NumPad1	Rewind
NumPad2	Fast-Forward
NumPad3	Record

NumPad4	Loop Playback toggle
NumPad5	Loop Record toggle
NumPad6	Quick Punch toggle
NumPad7	Metronome toggle
NumPad8	Count off toggle
NumPad9	MIDI Merge
NumPadEnter	Add Memory Location
NumPad*	Main Counter Select
NumPad/	Selection Counter Select
NumPad+	Nudge Forward
NumPad-	Nudge Backward
CtrlNumPad1	Transport
CtrlNumPad2	Session Setup
CtrlNumPad3	Main Counter
CtrlNumPad4	Automation Enable
CtrlNumPad5	Memory Locations
CtrlNumPad6	Machine Track Arming
CtrlNumPad7	Video Universe
CtrlNumPad8	Beat Detective

CtrlNumPad9	Video
AltNumPad0	Quantize (in Event Operations window)
AltNumPad1	Time Operations
AltNumPad2	Tempo Operations
AltNumPad3	Event Operations
AltNumPad4	Real-Time Properties
AltNumPad5	Elastic Properties
ShiftNumPad/	Link Timeline and Edit Selection
NumPad.(Number).	Recall Memory Location
NumPad.+	Add Windows Configuration
NumPad.(Number)+	Add Windows Configuration (specific number)
NumPad.(Number)*	Recall Windows Configuration

Chapter 12

Conclusion

My perspective of digital audio technology is better late than never. We, the music creators and Audio Engineers have always wanted a method to make things easier, faster, and sound better. When I say sound better, I am no way stating that the sound of digital audio is better than analog but stating that digital audio is clean and crisp while having the benefits of working less and letting the computer work for you. Everything in today's world has become digital based but at the same time, it's also eliminating jobs for humans. A majority of landmark recording studios has been shut down and I do blame digital technology for this because home studios are now affordable and convenient. Simple acoustic treatment can transform any space into a decent balanced room for mixing and recording vocals. Everything is at your fingertips, including shortcuts.

References

- **Avid-** The digital audio technology company, which acquired the Digidesign in 1995. Digidesign are the creators of the Pro Tools.
- **Wikipedia-** Encyclopedia
- **Soundonsound-** High technology Music Recording Publication since 1985

Glossary

AAC - Advanced Audio Coding (AAC) is a digital file format similar to MP3. Its probable best-known use is as the default encoding system used by Apple's iTunes.

Absolute Grid Mode - An editing mode that constrains movement and alignment of regions to precise increments on a user-defined grid.

A/D (analog-to-digital) - Analog-to-digital converters operate at various bit-rate resolutions and sampling rates, converting analog audio signals to digital audio signals.

AIFF (Audio Interchange File Format) -An audio file format invented by Apple Computer.

AFL (After Fader Listen) -This is a channel's level after it is attenuated or boosted by the fader setting. Audio tracks, Auxiliary inputs and Instrument tracks are AFL in Pro Tools

Analog – Analog is the opposite of digital. Any technology, such as vinyl records or clocks with hands and faces, that doesn't break everything down into binary code to work.

ATR (Audio Tape Recorder) - A tape recorder of any format, including a Digital Audio Tape (DAT) deck.

Audio Engineer - An audio engineer works with the technical aspects of sound during the processes of recording, mixing, and reproduction. Audio engineers often assist record producers and musicians to help give their work the sound they are hoping to achieve.

Audio Interface - Among most Pro Tools systems, the audio interface is a separate box that is attached with a special cable to the audio card, or to a USB or FireWire port. An audio interface typically has analog and digital audio inputs and outputs, and may be equipped with level meters, level controls, and other features.

Audio Region - A region of an audio file that is defined Nondestructively with pointers.

Auxiliary Inputs (Auxiliary Input Tracks) - In Pro Tools, these input channels are used for send and bus returns (input only). Sends are used for output routing

AVI (Audio Video Interleave) - AVI is a multimedia container format introduced by Microsoft as part of its Video for Windows technology. AVI files can contain both audio and video data in a file container that allows synchronous audio with-video playback.

Bit Depth - One of two main specifications that define digital audio quality (the other is sample rate). Bit depth determines the maximum dynamic range possible in an audio file. Also called bit-resolution or bit-rate.

Bounce to Disk - Mixing a segment of audio (or an entire session) internally to disk, without leaving the digital domain. Bit rate, dithering options, and other parameters are provided by Pro Tools for bouncing to disk.

Broadcast WAV File (BWF) - A variation of Microsoft's .WAV audio file format that contains

Additional data on the title, origination, date, and creation time of the audio content not included

In the standard .WAV file format. An important feature of BWFs is their support of time

Stamping. Time stamping allows files to be moved from one session to another and easily aligned to their original point in time.

Bus (noun) - An internal routing path.

Bus (verb) - To route one or more signals to one or more destinations (either internal or external).

Clipping - clipping indicator the LED at the top of each channel meter indicates a level may have run out of headroom, and is approaching clipping.

Clock Reference - Common "speed" reference, which various devices can use to establish synchronization during playback and recording.

Conductor Rulers - A ruler that can show session data. There are three types of Conductor rulers, called Tempo, Meter, and Markers rulers.

Crossfade - Function for fading out from one region as you fade in to another region. Crossfade duration is user-selectable from within the Edit window. As with fades, portions of audio for which the crossfade function has been applied are stored in the session's Fade Files folder.
D/A (digital-to-analog) - Digital-to-analog converters operate at various bit-rate resolutions and sampling rates, converting digital audio signals to analog audio signals.
DAE (Digidesign Audio Engine) - Digidesign's real-time operating system that provides the core functionality of hard disk recording, digital signal processing, mix automation, and MIDI required by Pro Tools and other Digidesign products.
DAW – Digital Audio Workshops used for the production of music, radio, television, podcasts, multimedia and nearly any other situation where complex recorded audio is needed, for example; Pro-tools, Logic, Reason, Etc.
Digital – Sometimes referred to as digital audio is a method of representing sound as numerical values. This differs from analog media such as magnetic tapes or vinyl records for instance where the sound is stored in a physical form. In the case of cassette tapes, this information is stored magnetically.
Dither - "Noise" added to an audio signal when down-sampling bit rates. Designed to create a smoother transition at lower amplitudes.
Drop frame - Refers to a variance of SMPTE/EBU time code for NTSC color video (29.97 fps) that omits two frames (frames "0" and "1") every minute except for every tenth minute.
DSP (Digital Signal Processing) - In audio terms, DSP refers to manipulation of digital audio, everything from reverberation to changes in level.
Elastic Audio - Feature in Pro Tools 7.4 and higher that lets you quickly and easily transpose, tempo conform, and beat match audio to the session's Tempo ruler.
Fade - A selection in which the volume rises or falls, typically from or to ∞. See also crossfade. **FireWire** - A high-speed peripheral standard capable of transferring data. FireWire is commonly used for digital audio and video devices, as well as external hard drives and other high-speed peripherals.

FLAC – Free Lossless Audio Codec
a file format for audio data compression that does not remove information from the audio stream, as MP3, AAC and Vorbis.
Frames per second (fps) - Number of frames that elapse per second,
as defined by the four SMPTE/EBU Time Code fps standards. These
include: • 24 fps (for film applications) • 25 fps (the PAL/SECAM
video standard) • 29.97 fps (the NTSC color video standard) • 30 fps
(the NTSC black and white video standard)
Grid Mode - Used to align regions in tracks to the grid or between
Grid boundaries. See also Absolute Grid mode and Relative Grid
mode
Groups - Linked tracks in which an action in one of the tracks is mir-
rored in all tracks in the group. Groups can be created separately or
linked between the Mix and Edit windows.
Headroom - Amount of remaining gain available for a given signal
before the onset of distortion in analog systems or clipping in digital
systems.
ILok (iLok USB Smart Key) - Portable, cross-platform USB device
for authorizing plug-ins and software options for Pro Tools systems.
For more information, see the iLok Usage Guide.pdf.
Inactive - Items that have been turned off in Pro Tools to free up or
conserve DSP. For example, when a track, send, or plug-in is inactive,
its name appears in italics and the item is silent.
Latency - Typically refers to the time it takes for an input signal to be
passed to the output and generally controlled by the hardware buffer
size.
Loop Sync - A dedicated clock signal for synchronizing multiple Pro
Tools|HD audio interfaces together. Loop Sync uses a Word Clock
signal based on sampling rates of either 44.1 kHz or 48 kHz. As sample rates increase in the system, Loop Sync continues to operate at a
base rate of 44.1 kHz or 48 kHz, depending upon the higher rate.
Loop Sync should be used only to chain multiple Pro Tools|HD pe-
ripherals together
Marker - Memory Location referenced on a timeline, typically used
to store locations to important points in a session.

Master – A stereo 2 track of your final recording presented in a master format for duplication purposes in either an analog ½" tape or high quality digital format.

Master (device) - "Lead" machine or Pro Tools system in a synchronized machine arrangement. Slaves follow masters. There can only be one master at any given time.

Master Fader - track Governs the overall signal level of one or more audio, auxiliary input, or instrument tracks.

Media file - A file that contains actual audio, video, or graphics data. It also contains a variety of metadata (such as file name and format).

Memory Location - Pro Tools supports up to 999 Memory Locations, which can include markers, Edit selections, record and play ranges, track settings, and other data. They can be viewed and sorted in the Memory Locations window, from which they can also be accessed.

Metadata - Media files, session files, and other types of files contain their own sets of metadata, which include general types of data such as file name, creation date, and file size. Metadata varies with file type, format, and kind.

MIDI (Musical Instrument Digital Interface) - A communication protocol designed to allow equipped instruments such as synthesizers to intercommunicate for control and playback purposes. Information transmitted over MIDI includes note-ones, note-offs, key velocity, pitch bend, and other performance data. Connections are made using cables equipped with 5-pin DIN connectors.

MIDI Editor Window - Dedicated editor window for in-depth editing of MIDI notes and controller data.

MIDI Event List - Pro Tools window that shows the contents of a MIDI track in a column, for easy editing of individual MIDI events.

Mix Engineer - A mix engineer or mixing engineer is a person responsible for combining ("mixing") the different sonic elements of a piece of recorded music (vocals, instruments, effects etc.) into a final version of a song (also known as "final mix" or "mix down").

MP3 - Mpeg-2 audio layer III or Mpeg Layer 3 compression to make the file size smaller.

MTC (MIDI Time Code) - Non-SMPTE form of time code that is used by some devices (including Pro Tools) to send and receive timing information.

Non-destructive Editing - Leaves audio files intact. As you edit audio within Pro Tools, you are only editing the regions, or "pointers," to audio files that are stored on the hard drive, unless you explicitly choose destructive modes (during recording, or when using AudioSuite processing).

Non-destructive Recording - Non-destructive recording is a method of recording where new takes or tracks are recorded while preserving the previously recorded tracks.

Non-Drop Frame - Time code that is not in dropframe format. In North America, the standard format outside of color video production or postproduction is typically 29.97 fps non-drop frame.

Non-linear Editing - Non-linear editing/recording typically refers to computer-based systems that allow any parts of the recording to be played back in any order with no gaps. Conventional tape is referred to as linear, because the material can only play back in the order in which it was recorded. Describes digital recording systems that allow any parts of the recording to be played
Back in any order with no gaps. Conventional tape is referred to as linear, because the material can only play back in the order in which it was recorded.

Notation - Traditional musical transcription, often used to visualize the composition process and provide written parts for musicians to play

NTSC - Video standards developed by the National Television Standards Committee. NTSC color video runs at 29.97 frames per second; NTSC black and white video runs at 30 fps. Used primarily in North and South America and Japan.

Nyquist frequency - The highest audio frequency that can be accurately sampled, equivalent to one-half of the sampling frequency. The Nyquist sampling theorem showed that the sampling rate must be at least twice the highest frequency present in the sample in order to accurately reconstruct the original signal.

Offline (media) - Not connected to or directly accessible by a computer.

Offline (synchronization) - Not under the control of (or controlling) another device for synchronized playback or recording.

Online (media) - Connected to and directly accessible by a computer.

Online (synchronization) - Controlled by (or controlling) another device for synchronized playback or recording.

Overdubs - Record (additional sounds) on an existing recording.

Peak Indicator - Indicator light designed to warn of the possibility of clipping, which illuminates as a device's input reaches a preset degree of headroom.

Plug-in – Also known as an add-in, add-on, or extension is a software component that adds a specific feature to an existing computer program. When a program supports plug-ins, it enables customization which enhances your primary or another software.

Post-fader - Output from a track (typically a send) that is governed by the channel's fader setting.

Post-roll - Adjustable time for playback to continue beyond the current playback or recording of a selection.

PRE - Digidesign's remote controllable 8-channel microphone preamp. Features eight discrete, matched transistor hybrid mic-preamp circuits and offers a pristine signal path designed specifically for the Pro Tools environment, but can also be utilized as a stand-alone microphone preamp.

Preamp - In recording studio terminology, a circuit designed to boost relatively low signal levels, such as a microphone output, up to standard line levels of −10 dB or +4 db. Many Pro Tools LE systems are equipped with microphone preamps. See also PRE.

Pre-fader - Output from a track (typically a send) that is independent of the channel's fader setting.

Pre-roll - Adjustable time that precedes the playback or recording of a selection

Pro Tools - Digital audio workstation software that lets you record, arrange, compose, edit, mix, and master professional quality audio and MIDI for music, video, film, and multimedia.

Pro Tools Creative Collection - Set of RTAS® (Real Time Audi-oSuite) instruments and effects plugins included with Pro Tools. For more information, see the Creative Collection Plug-ins.pdf.

Pull Up/Pull Down - Refers to the deliberate "miscalibration" of the audio or video sample rate clock (the audio pitch) in order to compensate for a speed change elsewhere in the production chain. The usual situation in which these rates are encountered is when film footage (at 24 fps) is

Transferred to color NTSC-standard video tape (at 29.97 fps). Video Pull Up/Pull Down is supported in Pro Tools 5.3.1 and higher.

Quantization – The process of aligning a set of musical notes to a precise setting or tempo map. This results in notes being set on beats and on exact fractions of beats. The most frequent application of quantization in this context lies within MIDI application software or hardware.

QuickTime - Apple's system extension for control of time-based events, such as digitized video

Movies and digitized sound.

Region - Within Pro Tools, a "pointer" to a particular track selection or file. Regions can be dragged from the Region List, or a DigiBase browser, to a track.

Relative Grid Mode - An editing mode that constrains movement and alignment of regions to precise increments on a user-defined grid while allowing the region to maintain an offset relative to that grid point.

Sample Rate – A sample rate is the number of samples of a sound that are taken per second to represent the event digitally. The more samples taken per second, the more accurate the digital representation of the sound can be. For example, the current sample rate for CD-quality audio is 44,100 samples per second. This sample rate can accurately reproduce the audio frequencies from 20 hertz up to 20,500 hertz, covering the full range of human hearing.

Send - An adjustable additional track output, which can be routed to an internal bus for effects processing, monitoring, and sub mixing.

Shortcuts - Pro Tools keyboard and Right-click shortcuts that give you fast access to a wide variety of tasks.

SMPTE Time code - is a set of cooperating standards to label individual frames of video or film with a time code defined by the Society of Motion Picture and Television in the SMPTE 12M specification. SMPTE revised the standard in 2008, turning it into a two-part document: SMPTE 12M-1 and SMPTE 12M-2, including new explanations and clarifications. Time codes are added to film, video or audio material, and have also been adapted to synchronize music. They provide a time reference for editing, synchronization and identification. Time code is a form of media metadata. The invention of time code made modern videotape editing possible, and led eventually to the creation of non-linear editing systems

Spot - Audio post production process of aligning audio events to visual events. In Pro Tools, Spot mode lets spot regions to particular time code events.

Standard MIDI File - Universal format that can be read by virtually any software that reads MIDI. Type 0 is a single line sequence, type 1 is multitrack.

Sub mix - Routing multiple audio sources to an Auxiliary Input for monitor mixes, bus-master control over levels, and shared effects processing.

Subgroup - Refers to a console's output busses (stems, cue stems) in standard audio terminology.

TDM (Time Division Multiplexing) - A technology that employs a networked bus of DSP chips that supply the processing power for Pro Tools|HD systems. This bus is called the Digidesign TDM Bus™.

TDM plug-in - Digidesign's proprietary real-time, nondestructive plug-in format for Pro Tools|HD systems.

Tick-based - Editing mode in which audio and MIDI regions and MIDI notes are snapped to the nearest MIDI tick value. Switchable in Pro Tools on a per-track basis.

Timeline insertion point - Location on the timeline corresponding to the cursor point, and the point from where playback or recording begins.

Universe window - Provides a visual overview of all tracks in a session, and can be used to quickly navigate to any location in a session.
USB (Universal Serial Bus) - A high-speed peripheral standard. USB 1.0 is capable of transferring data at up to 12 Mb/sec, whereas USB 2.0 is capable of transferring up to 480 Mb/sec. USB is used for many audio and video peripheral devices.
Velocity - MIDI data parameter that describes how fast or hard a key is struck and controls the volume of MIDI note playback.
Virtual Instrument - Software-based MIDI instrument, often in plug-in form, that is used to replace or augment hardware-based synthesizers, samplers and drum machines.
Voices - With a Pro Tools system refers to the number of channels that can be played back simultaneously.
VST – Known as Virtual Studio Technology (VST) is a software interface that integrates software audio synthesizer and effect plugins with audio editors and recording systems. VST and similar technologies use digital signal processing to simulate traditional recording studio hardware in software.
WAV - This is Microsoft's Audio File Format. Can be read by Pro Tools on both Windows and Mac platforms.
WaveDriver - The Digidesign WaveDriver is a single-client, multi-channel sound driver that allows third-party audio programs that support the WaveDriver MME (Multimedia Extensions) standard to play back through Digidesign-qualified audio interfaces. For more information, see the Windows Audio Drivers Guide.pdf.
Waveform - Means of visually representing a sound. When sound regions are imported into the Pro Tools Edit window, they can be viewed in Waveform view. Preview waveforms can also be viewed in DigiBase browsers
Word Clock - Many professional digital audio products—including open-reel multitrack tape recorders, digital mixing consoles, and digital recorders—have Word Clock (1x sample rate) connectors. Word Clock allows Word Clock-compatible devices to send or receive external clock information that controls the sample rate, which in turn (where applicable) controls the play and record speed.

Zero crossing - Point at which a wave's amplitude crosses the center-line of the waveform display. Typically, a good spot to edit a sound file is at zero crossings, to avoid unwanted artifacts.

Zoom - Function used to view waveform displays within the Edit window with greater detail or more data.

* 9 7 8 0 5 7 8 1 7 9 0 5 6 *